MACMILLAN · FIELD · GUIDES

ROCKS & ·MINERALS·

The authors

DAVID WRIGHT, Ph.D., is Lecturer in Geology at Kingston Polytechnic, Surrey, England. He specializes in Petrology.

PAT BELL is a consultant geologist and sedimentologist whose speciality is the Mediterranean region, including North Africa, the Middle East and North Sea.

Acknowledgements

The authors would like to thank Rob Ravenhill for his help in sorting out the many rock specimens.

Photographs

Pat Bell 35, 37, 41; Institute of Geological Sciences 27; Natural Science Photos 21 (top); RIDA/David Bayliss 28, 31, 33; David Wright 8, 16, 21 (bottom).

Identification plates and cover by David Johnson.

Inset photographs on pages 125, 127, 129, 135, 137, 139, 141 by Linda Parry.

Illustrations by The Hayward Art Group.

This Edition Published 1985 by Macmillan Publishing Company,
a division of Macmillan, Inc.

Macmillan Publishing Company
866 Third Avenue, New York, N.Y. 10022
Collier Macmillan Canada, Inc.

Library of Congress Cataloging in Publication Data

Bell, Pat.
Rocks and minerals.
(Macmillan field guides)
Includes index.
1. Rocks—Identification. 2. Rocks—Collectors and collecting. 3. Mineralogy, Determinative. 4. Mineralogy—Collectors and collecting. I. Wright, David. II. Title. III. Series.
QE431.5.B43 1985 552 84-17578

ISBN 0-02-079640-4

10 9 8 7 6 5 4 3 2

Printed in Italy

9/87

MACMILLAN · FIELD · GUIDES

ROCKS & ·MINERALS·

PAT BELL · DAVID WRIGHT

Collier Books
Macmillan Publishing Company
New York

Foreword

The rocks which make up the Earth influence our lives more than many of us imagine. Rocks determine the form of the landscape, and give rise to the loose covering of soil. In cliff faces, mountain sides, and river valleys, for example, we can find naturally exposed rocks, while man-made exposures, such as cuttings and quarries, allow us access to the rocks which would otherwise be hidden from view. But, what is a rock? To a geologist (a scientist who studies the Earth), a rock is any naturally formed aggregate of minerals, whether consolidated or not, and a mineral can be defined as a homogeneous, solid substance, with a fixed chemical composition, formed by the inorganic processes of nature. Thus, even a soft clay is a rock. This book is for the growing army of people who are realizing that the study of rocks and minerals is a fascinating and absorbing pastime.

Contents

Introduction

This book will help you collect and identify rocks and minerals, but is also intended to help you understand or 'read' the landscape, so that you can relate the underlying rocks to the land forms you see around you. The book describes how the three main groups of rocks arose and how their observable features can be used to identify them, and give us some clue of the processes that built them. The colour plates illustrate the more common rock types, together with a number of their main constituent minerals. The accompanying descriptive text provides information on their composition, formation and mode of occurrence.

How to use the book

The introductory chapters give you a clearer idea of what you are looking at and how best to go about collecting effectively and safely. Once you have obtained your sample, you may be able to identify it by its occurrence in the field. If not, refer to the first key (*see* page 49) and establish whether you are dealing with an igneous, sedimentary, or metamorphic rock. Next, refer to the relevant key for the rock type that you think you have collected. This key should enable you to turn to the correct pages of the book where you can confirm your identification using the colour plates and descriptions. These will also help you to identify the minerals of which rock is composed. It may be possible simply to compare your sample by flicking through the colour plates.

Conservation and safety

Over-collecting has already destroyed some unique localities and, in addition, many land-owners are concerned about people visiting sites on their land. Please try to observe the following simple rules:

1 shut all gates;
2 leave no litter;
3 follow any local bylaws;
4 always try to obtain permission before entering private land;
5 do not leave rock fragments strewn over fields or roads;
6 avoid disturbing wildlife or farm animals.

Beware of unstable and insecure rock faces or cliffs when you are looking at rocks at the coast or in quarries – and don't forget about local tide conditions! Try not to hammer indiscriminately at every rock you see and don't collect more than you need. When you are hammering, especially on hard rocks, protect your eyes from flying rock fragments by wearing protective goggles.

Rock and mineral collections

All specimens should be carefully numbered, and details of the locality from which the specimens were collected, together with associated measurements and information, should be recorded in a notebook. Wash specimens to remove any soil and dust, and dry before storing.

Equipment

An item of equipment which is usually essential is a geological hammer. Those used most commonly are the 1 pound and 2.5 pound (0.5 kilogram and 1 kilogram) versions. To extract specimens carefully, you may need to use one or more cold chisels. A broad bolster chisel is ideal for collecting specimens which split easily. Plastic bags, waterproof marker pens and a notebook in which to record details are also important.

To examine rocks, and possibly for identifying minerals, you need a hand-lens. You should also carry a piece of unglazed porcelain to use as a streak plate (*see* page 16) and a pocket knife and a piece of window glass for assessing hardness (*see* page 17). To test whether or not a rock contains calcium carbonate, it is useful to have some white vinegar. (Rocks containing calcite will fizz on contact with vinegar.) Carry a camera, too, because you can often photograph geological phenomena rather than disfiguring them by collecting. If you wish to investigate the structure of the rocks by measuring the angle at which the beds of the rocks are lying you will need a compass clinometer. The compass will give you the direction of strike of the rocks, while the clinometer will give you the angle of dip (*see* page 10). Compass clinometers are expensive, but you can use any magnetic compass and a home-made clinometer. The clinometer consists of a perspex protractor and a weighted thread. The straight line between the two 90 degree marks is used as the measuring line. This is aligned with a flat bedding plane of the rock surface. The deviation of the weighted thread from the central line is measured from the protractor scale to record the dip of the rock.

Wear stout, comfortable, and, preferably, waterproof boots and thick socks. It is wise to wear proper walking trousers of a suitable material, or even breeches or shorts. Wear several light but warm shirts and sweaters which trap plenty of air but allow you to adjust your dress as the conditions change. Even in sunshine, take a lightweight waterproof.

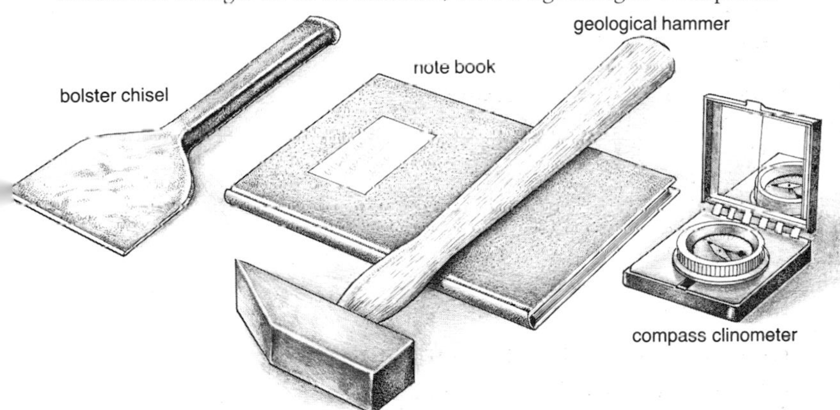

Basic equipment required for geological field work. In addition, you will need a rucksack for carrying your equipment and any samples that you collect.

Fieldwork

To a geologist, the 'field' is anywhere that rocks may be observed in situ. Geological information is gained initially in the field, and then backed up with laboratory studies. How you go about field work will be determined by what you are trying to discover. You may just wish to visit a single outcrop or quarry and notice the relationships between certain rocks and collect some specimens, or you may decide to carry out much more detailed work by studying many outcrops and quarries.

Making a start

First decide what you want to do, and how much time is available. You may be planning to spend several weeks in an area, or you may only wish to spend a few hours at an outcrop. Make sure, too, that you have the right equipment and suitable clothing and footwear. You will need to know exactly where you are, so always carry an accurate topographic map at a scale of 1:50000 or, for even greater accuracy, 1:25000. Examine the map carefully and try to locate quarries and large outcrops, because it is best to visit good exposures first so that you can then recognize poorly exposed rocks later. It is a good idea to keep your map on a clipboard with a waterproof cover together with a notebook for taking field notes.

When you arrive at an outcrop look for any obvious features such as bedding, and whether the beds have been folded as shown here. Note also in this outcrop the joints arranged radially around the fold. The best collecting sites are quarries, cliffs and cuttings.

Field notes

Good field notes are vital because they may be your only record of what you have found at a particular site, and you or someone else may need to refer to them if you return to the site. Your notes should be clear and concise, with sketches and photographs of outcrops and any other relevant features. You should keep a careful record of each photograph taken, with a simple diagram noting the direction of view and other important features. It is surprising how easy it is to end up with a series of rock photographs which mean very little unless they are accompanied by accurate notes.

When you are writing field notes it is a good idea to write them as though someone else is going to use them later to find and examine an outcrop. Each locality (an outcrop or any other place that you obtained relevant information) should be carefully located using co-ordinates – if you have a map – or by compass bearings on prominent landmarks. To take a bearing, sight your compass across to the first landmark, align the compass to north and take a reading; then repeat for a second landmark approximately 90 degrees away from the first one. Anyone will be able to use these bearings and, by adding 180 degrees to each, may draw lines on the map from the landmarks and locate your position fairly accurately. Don't forget that you should always allow for the deviation of magnetic north from north as indicated on the map.

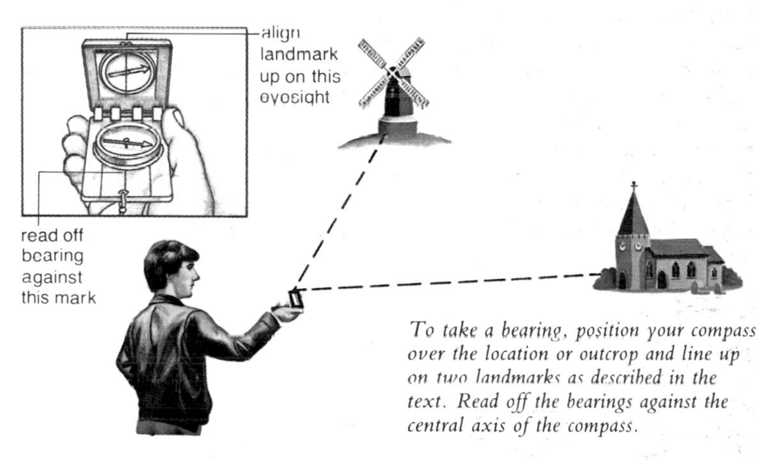

To take a bearing, position your compass over the location or outcrop and line up on two landmarks as described in the text. Read off the bearings against the central axis of the compass.

Collecting samples

This should always be done sparingly, taking care not to damage the outcrop too much. A specimen should be representative of the outcrop but, if the outcrop is very variable, then you should take a series of small samples. The size of the sample will depend upon the grain size of the rock, but do not take unnecessarily large samples, especially if you have to carry them a long way! Fist-sized pieces are usually adequate. Write a number on the rock sample itself and place it in a similarly numbered bag (if you cannot write on the rock, put a numbered piece of paper in the bag). This double coding is used because numbers can rub off quite easily. Note the number and other details of the samples in your field notebook.

AT THE OUTCROP OR QUARRY The first thing to do when you arrive at a site is to take a step back, look at the exposure as a whole, and work out its general features. Look for features such as bedding in sedimentary rocks or, possibly, flow banding which can indicate igneous rocks. Look also for veins and open fractures where minerals (particularly quartz or calcite) often form large, well-formed crystals. If you plan to examine the geology rather than just collect specimens you should certainly try to establish the dip and strike of the rocks using your compass clinometer (*see* below). In this way, you will be able to determine the structural features of the rocks and how they relate to one another and to their formation. The dip is the amount by which the bedding plane (*see* page 30) is inclined to the horizontal. The strike is a horizontal line drawn on the dipping bedding plane. (A simple analogy is a roof, where the strike is the ridge or any line parallel to it, and the dip is the slope of the roof at right-angles to the strike.) Find the line along which the dip is the greatest – you could drip water on to the bed and it will run down the steepest path. Measure the inclination and the approximate direction of dip. Draw a line at right angles to the dip and measure the orientation of this line in the same manner as you took the bearing (*see* page 9). Thus, if the rocks have been folded, these measurements will help you to determine the structure of the folding. If, however, you are only planning to collect and identify rock and mineral specimens, you need not necessarily take any of these measurements.

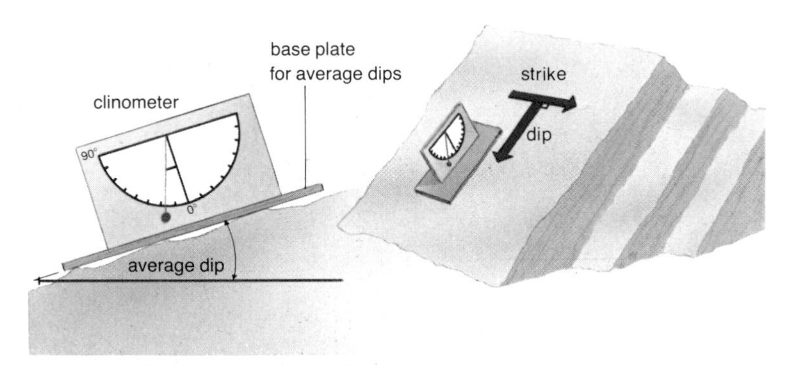

Taking dip measurements (left). On a smooth bedding plane only the greatest angle measured represents true dip. On irregular surfaces use your notebook as a base plate. On the right is shown the relationship between dip and strike.

Having obtained some idea of the general rock type, move closer to the outcrop and select some samples. If you observed some kind of boundary or change, then you should sample both sides of the boundary. You should collect 'in situ' samples to be certain of their relative positions in the outcrop. Choose a fairly solid corner and hit it firmly several times with your hammer. You may need to use your chisel to extract a sample where no suitable outjutting of the rock occurs. Examine the freshly broken surface of the rock. Use your hand-lens by holding the lens close to your eye and moving the rock (rather than the lens) to bring it into focus. By using this method you get a good field of view. With the lens you should be able to see the individual rock components and their mutual relationships.

Test individual minerals for hardness by scratching them with your knife. Wipe the mineral with your finger and look for a scratch (a mark can often be caused where the metal of the knife blade has been left on the mineral, so beware). If no scratch is seen, try scratching the flat of your blade with the mineral. If both tests appear to cause a scratch, or neither does, then the mineral is probably of a similar hardness to your knife. Use the streak plate by passing the mineral over the surface of it with about as much pressure as you would use when writing. Blow the rock dust off the plate and examine the streak (if any). A white streak is often difficult to see. Dilute hydrochloric acid or vinegar can be used to identify a carbonate mineral or to see if the rock has a calcareous matrix (*see* page 126). Put a single drop of acid on the surface of the mineral or rock, if it effervesces vigorously then there is probably a lot of calcite. If the sample effervesces only slightly look at the area under the acid drop; is the whole area bubbling or just the matrix between the grains (as in the case of a calcareous sandstone – *see* page 126)? Use the flow diagrams and the illustrations to identify the rock type. (Individual rocks and minerals and their appearances are described between pages 56 and 181).

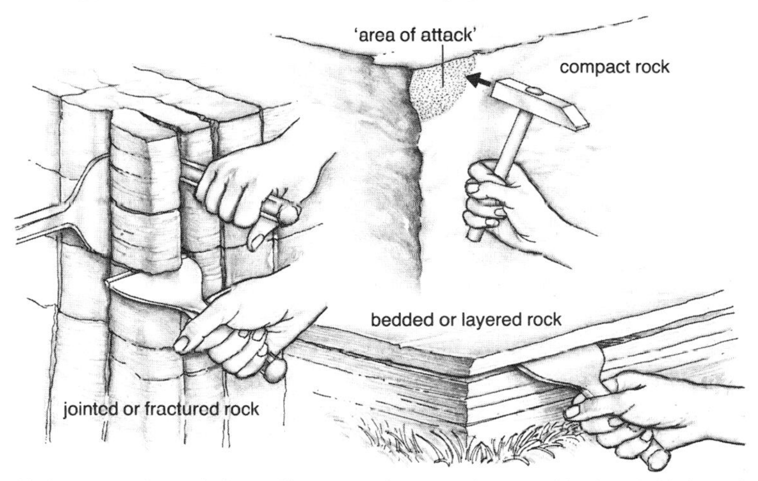

Various extraction techniques. On some rock types using a combination of chisels, and techniques, may be necessary. Try not to hammer indiscriminately at every rock you see.

Collecting fossils

Note should be taken of the rock type in which these are found. If possible, sample loose rocks at the foot of the outcrop, as often dozens of rocks have to be broken open before even one fossil is found. Fossils usually occur on bedding surfaces. Prise rocks open along these planes using your hammer and chisel or, in the case of soft shales, your penknife. If you must collect 'in situ' fossils then take great care not to damage the outcrop.

How to use geological maps

Printed geological maps can provide a lot of valuable information, but you should be aware of their limitations. They are drawn from the information which the geologist can glean from outcrops of the rocks. The bright, solid colours on printed maps give the impression that, if you went to a particular place on the map, you would find a particular rock. This is not always the case, because exposures of rocks may well be widely scattered, with large areas covered with soil or even towns and cities. But, if you look for quarries and large exposures which will be marked on the map, then you should find the rocks you are looking for.

Hills and valleys can make a structurally simple area appear complex. On most geological maps there is a cross-section which gives you an idea of the structure and the relationships of the various rock types to one another. Dip and strike symbols will also help to elucidate the structure. On the whole, however, some imagination is needed to visualize the hills and valleys in three dimensions so that you can then try to interpret the outcrop patterns accordingly.

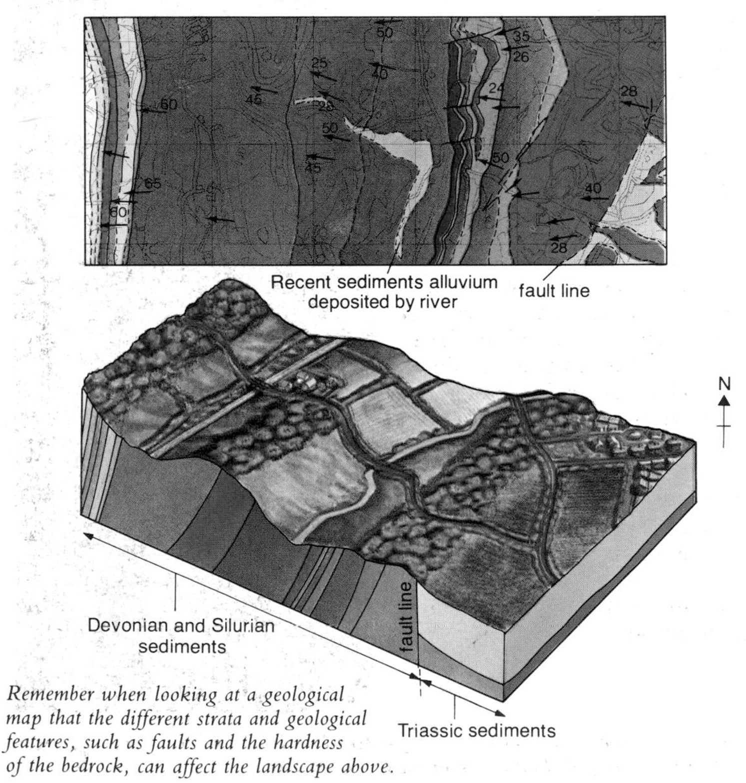

Remember when looking at a geological map that the different strata and geological features, such as faults and the hardness of the bedrock, can affect the landscape above.

Geological timescale

The names of rock units and their divisions are often complicated, and are based on a variety of factors, such as the fossil content, the age, or type of rocks. The geological timescale is not universally agreed upon. The time boundaries may vary with opinion and names change with the area to which they are applied. For example, in Europe the Carboniferous is usually considered to be a single period while, in North America, it is divided into the Mississippian and Pennsylvanian systems.

On geological maps there is usually a column which represents the ages and relationships of the various rocks found in the area. In general, the divisions which represent the greatest time span are called **eons** which, in turn, are subdivided into successively shorter time units, such as **eras**, **periods**, **epochs**, and **ages**. Rock units may also be given names, although they may comprise several beds of different rock types and may not be confined to a particular division of time. This type of arrangement is referred to, therefore, as a **diachronous** classification. There is an internationally agreed arrangement of rock units into divisions of decreasing size from **group**, through **formation** and **member**, to **bed**. All these divisions and variations may seem confusing, but in terms of a single geological map the divisions have generally been applied to simplify the situation.

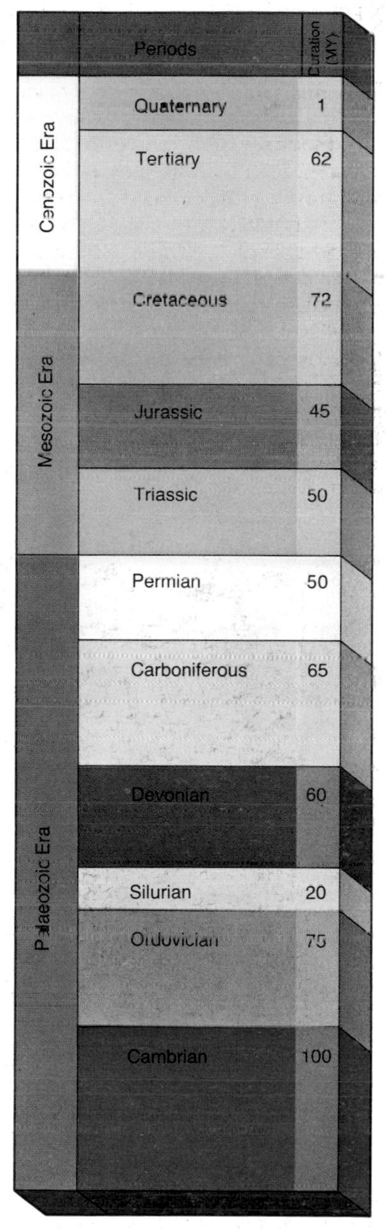

The geological time scale showing the major periods. The Cambrian Period is preceded by an even longer period called the Precambrian.

Minerals

All rocks are composed of aggregates of different **minerals**. A mineral is any naturally occurring solid substance with a more-or-less definite chemical composition and physical properties. Every mineral contains one or more of the many naturally occurring chemical elements. If it is made of just one element, such as copper, or diamond which is made of carbon, it is called a **native element**. More commonly, however, minerals consist of **chemical compounds** where two or more elements are combined. Minerals usually occur in some kind of **crystal** form which reflects the way in which the atoms that make up the mineral are arranged into orderly geometrical patterns. When they are allowed to grow freely, minerals have regular shapes with smooth, plane (flat), faces, while those which grow as aggregates are less perfect. Well-formed crystals of a particular mineral always assume a characteristic shape which is diagnostic for that mineral.

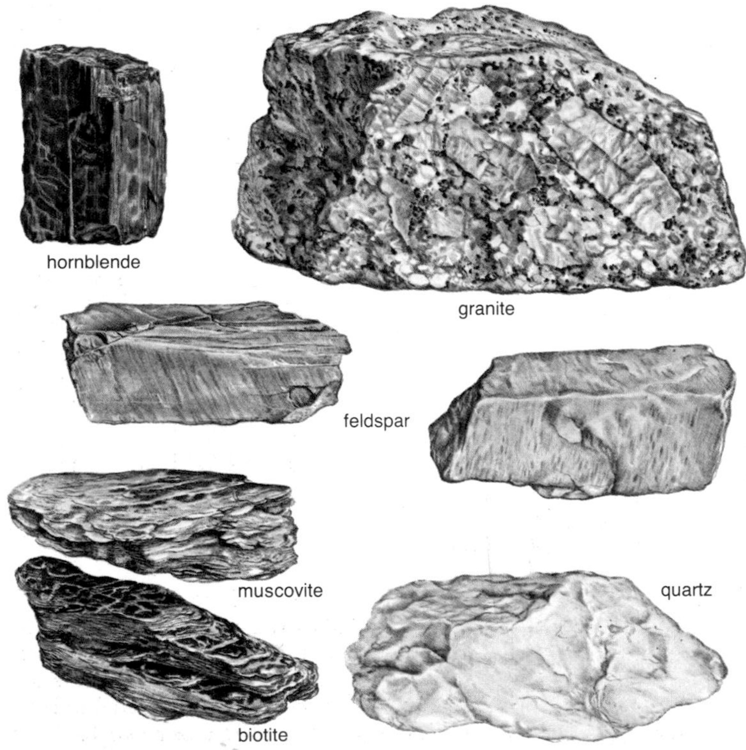

Above: An example of granite with some of its constituent minerals. Opposite: The seven crystal systems: shown here is a diagrammatic representation of the crystal axes, with an example of a crystal from each.

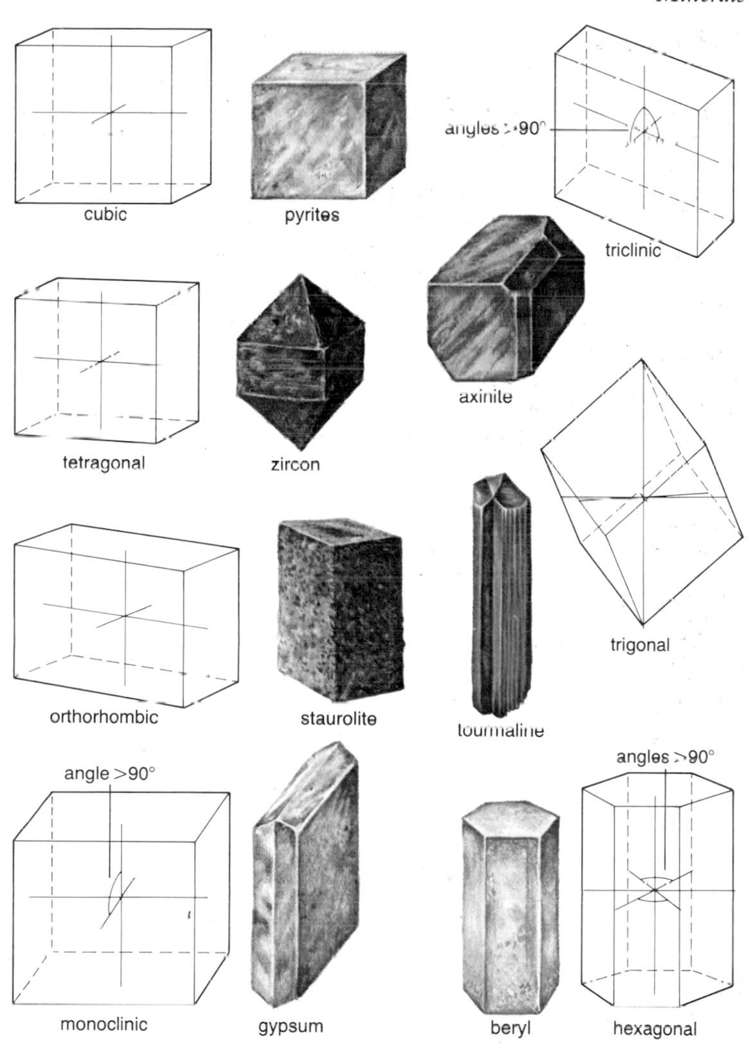

The symmetrical and repetitive arrangement of the faces of a crystal reflects the arrangement of its atoms, and allows us to divide all crystals into seven main **crystal systems** known as cubic, tetragonal, hexagonal, trigonal, orthorhombic, monoclinic, and triclinic. These systems are defined on the basis of axes about which the crystal can be rotated and still show symmetry. For example, a tetragonal crystal has one four-fold axis of rotation. In other words, if you rotate the crystal about this axis, you will see an identical face four times. But each face may develop differently, even though the angles between the faces remain consistent.

The different development of the faces gives a crystal its characteristic **habit**. For example, the crystal may be flat and slabby, when it is called tabular, or it may be elongated and referred to as prismatic. The arrangement of the atoms within a mineral, together with its chemical composition, gives the mineral a number of physical properties which can be used to help identify that mineral in the field.

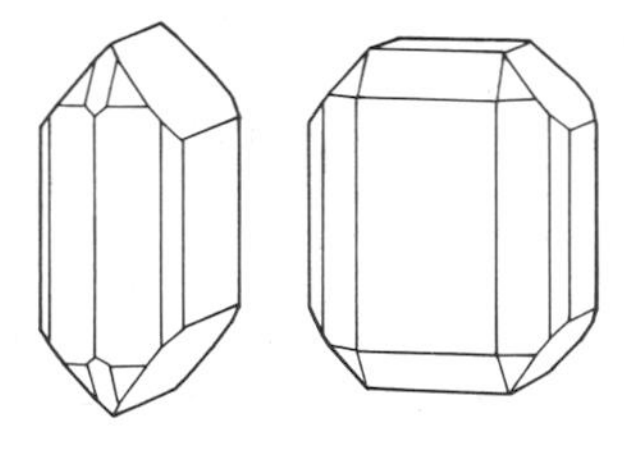

Tabular crystal habit shown by mica (above), and examples of prismatic crystals shown by many of the silicate minerals (right).

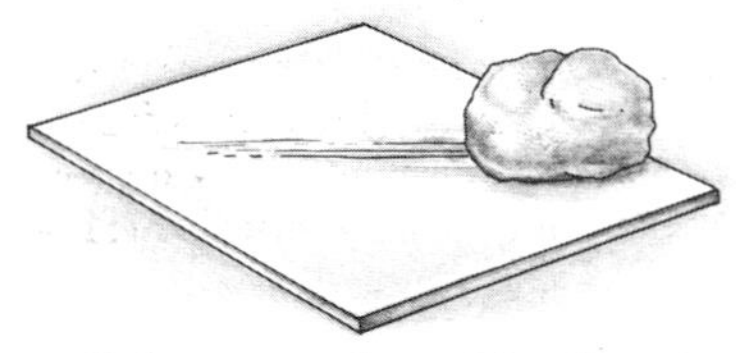

Left: A mineral being drawn across a streak plate (a piece of unglazed porcelain) to produce a smear or streak of the powdered mineral. This may be diagnostic for minerals which produce a streak.

Colour can be misleading in a weathered specimen of a mineral because the colour that you see may merely be that of the products of weathering. A far better indication of the mineral's colour is that which can be seen in powdered form. In the field, this can be obtained by scratching the mineral on a piece of unglazed porcelain to give a **streak**.

The surface of a mineral will reflect light to different degrees and in different ways. This is called the mineral's **lustre**. For example, a mineral may have a metallic, glassy or vitreous, pearly, resinous, silky, or dull lustre.

Examples of common lustres shown by various minerals. A very well-developed metallic lustre exhibited by the sulphide mineral, galena (left), and a glassy or vitreous lustre shown by the pink variety of the common silicate mineral, quartz (right). Due to its colour, this variety is called rose quartz.

The **specific gravity** of the mineral may be useful, too. This is defined as the ratio of the mass of a given volume of the mineral compared to the mass of the same volume of water. In the field, you can get a rough idea of this simply by 'hefting' the specimen in your hand. With experience, you will be able to tell whether the specimen feels 'heavy' or 'light' for its size.

Probably the most important single feature of a mineral for identification purposes is its **hardness**. In other words, you can determine whether or not the specimen will scratch or will be scratched by another object or mineral of known hardness. There is a list of ten minerals, belonging to the standard **Mohs' scale** of hardness, which can be numbered in order of increasing relative hardness: talc 1; gypsum 2; calcite 3; fluorite 4; apatite 5; orthoclase feldspar 6; quartz 7; topaz 8; corundum 9; and diamond 10. In other words, fluorite will scratch calcite but will be scratched by apatite. In the field it is unlikely that you will have the standard samples but, as a rough guide, your finger nail is about hardness 2½, a copper coin about 3, window glass 5½, and a steel knife blade about 6½.

Most minerals will split relatively easily in certain directions to yield smooth, parallel, often closely spaced planes called **cleavage planes**. These planes reflect the atomic structure of the mineral and are often parallel to crystal faces. Some minerals, such as quartz, do not cleave but **fracture** irregularly.

Biotite has one perfect cleavage formed of one set of parallel cleavage planes. It is easily split along these planes.

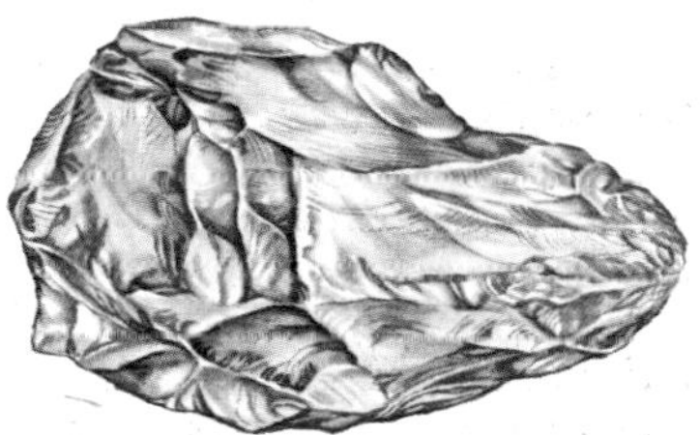

Quartz has no cleavage. Thus it does not split easily, but fractures unevenly, developing irregular fracture surfaces.

Minerals can be classified on the basis of their chemical composition into the following main groups: native elements (diamond), oxides (corundum, magnetite), sulphides (pyrite), halides (fluorite), carbonates (calcite), sulphates (gypsum), phosphates (apatite), and the main rock-forming minerals, the silicates. After the silicates, the carbonates are the next most important rock-forming group.

Essentially, the silicates are made up of units of four oxygen atoms surrounding a single, smaller silicon atom in a tetrahedral arrangement. Millions of these tetrahedra may then be linked together as groups of separate, compact, and dense units, such as in olivine or garnet, in chains such as in the pyroxenes, in rings such as in tourmaline and beryl, as sheets such as in the micas, or in a framework such as in the feldspars or quartz. Chain minerals usually have two cleavage planes, while sheet minerals show one strong cleavage plane, parallel to the layers of atoms.

Igneous rocks

Origin of igneous rocks

If you ever have occasion to visit a mine, particularly a deep one, you will detect a gradual rise in temperature the deeper down you go. Therefore, if it were possible to descend far enough into the Earth, depths would be reached where temperatures are high enough to partially or completely melt the surrounding rocks. This molten or liquid rock material, which also contains some dissolved gases, is called **magma**. When it erupts to flow over the Earth's surface, it is called **lava**. When a magma or lava cools and completely solidifies, an igneous rock is formed.

The majority of magmas contain between 40 and 75 per cent silica, which has been derived from the melting of rocks in which silicate minerals are the main constituents. Thus, the resulting igneous rocks are also predominantly composed of silicate minerals, such as olivine, pyroxenes, amphiboles, micas, feldspars, and quartz. It is possible to measure the temperatures of lavas as they erupt. Their temperatures range from less than 700°C (1300°F) to well over 1200°C (2200°F). The chemical composition of the lavas varies widely, too, but it also bears some relation to the temperature. In general, variations in the physical properties of the lavas (and the magmas) are consistent with chemical differences.

MAGMA This is produced wherever the temperature (and pressure) conditions cause rock to melt. This usually means that magma can only form deep below the Earth's surface, either within the lower part of the outermost zone of the Earth, the **crust**, or the middle zone, or **mantle**. The way in which a rock melts is an extremely complex process influenced not only by temperature and pressure, but also by the presence of gases such as steam and carbon dioxide. In addition, it is quite rare for a rock to melt completely. Most rocks are composed of several minerals, usually silicates, which all have different individual melting points. Thus, the complete melting of such a mixture may only be achieved over a particular temperature interval, say 200 to 300°C (390 to 570°F).

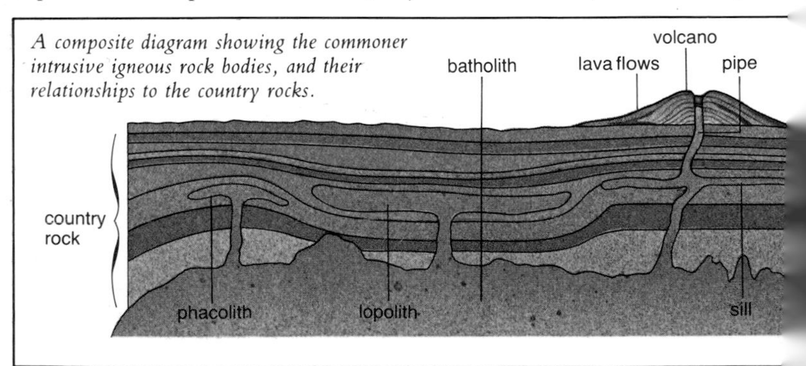

A composite diagram showing the commoner intrusive igneous rock bodies, and their relationships to the country rocks.

Consequently, it may be that the existing temperature may only induce melting of a proportion of the minerals present so that the rock is not completely melted. The first-formed magma will separate off from the unmelted rock and, furthermore, the composition of this magma depends on the part of the rock first melted. This process is known as **partial melting**.

Such processes occur at great depths within the Earth – up to 50 miles (80 kilometres) – but they can be duplicated in the laboratory.

Chemical composition

Magmas and lavas are chemically complex. A variety of elements and compounds, such as aluminium, iron, magnesium, calcium, sodium, and potassium, are found in naturally occurring lavas and magmas, and these may, in turn, be combined with oxygen to form oxide compounds. The most important compound, however, is silica, and its presence provides a ready method of classification. The percentage abundance of silica can be determined by chemical analyses of magmas and lavas, and the resultant igneous rocks. Magmas, lavas, and igneous rocks possessing between 45 and 52 per cent silica by volume are termed **basic**; those having between 52 and 66 per cent silica are **intermediate**, while those possessing more than 66 per cent silica are called **acid**. From basic to acid magmas and rocks, the iron and magnesium content decreases while the content of sodium and potassium increases.

Modes of occurrence of igneous rocks

Once formed, magma tends to rise – albeit very slowly – because it is less dense than the surrounding rocks from which the magma has been derived by partial melting. As the rocks melt they expand, and the pressures which result also tend to cause the magmas to rise. The magma is squeezed upwards, often into deep fractures within the rock which have been caused by this deep-seated pressure. The rise of magma may continue until it reaches the Earth's surface. Here, the lava is extruded on to the surface through volcanoes and forms lava flows. Thus, igneous rocks formed from magma which consolidated at the Earth's surface are

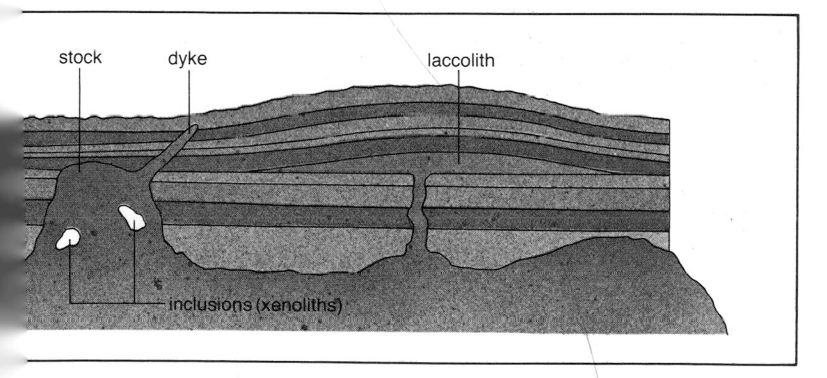

called **volcanic** or **extrusive igneous rocks**. In some cases, however, the magma may never reach the surface of the Earth and solidifies at some distance below. Such igneous rocks are termed **plutonic** or **intrusive igneous rocks**. We are still able to look at intrusive igneous rocks because the rocks which originally lay on top of them have been gradually stripped away by many millions of years of continual erosion by such processes as the action of wind, water and ice.

Extrusive igneous rocks

Volcanic rocks are generally found as lava flows of different shapes and sizes. The lava reaches the surface either through long, linear cracks called **fissures**, or through a single pipe, called a **conduit**. Fragments of lava and other debris gradually build up around this conduit, to produce a volcano. The products of volcanic activity include gases, liquids, and solids. Gases (steam and sulphur) are largely lost to the atmosphere. The liquid lava solidifies to produce extrusive igneous rocks.

If you have watched a film of a volcanic eruption or been fortunate enough to observe one yourself, you may have noticed that the volcanic explosions blast material up into the air. This material may be solid, or it may be liquid and solidifies as it falls through the air. Eventually, all this fragmented material falls back down to earth. The large blocks and 'bombs' descend very quickly, while the much finer ash takes much longer to settle. During the course of an eruption, and following subsequent eruptions, this **pyroclastic** material gradually builds up around the volcano, forming layers which eventually consolidate to form **pyroclastic rocks**. Such rocks are frequently associated with extrusive igneous rocks.

Lava flows vary greatly in appearance and character depending upon how **viscous** or 'runny' the lava is. The silica content controls the viscosity of lavas. The more silica a lava or magma contains, the more viscous it will be and the less easily it will flow. (To illustrate this, think of the way in which water, oil, and treacle flow. Water flows easily, oil slightly slower, while treacle flows slowest of all. Therefore, oil is more viscous than water, and treacle more viscous than oil.) Temperature also affects viscosity. To use oil as an example again, you will know how difficult it is to pour oil from a can on a cold day, but if the can is warmed, the oil will flow much more easily.

Basic lavas, therefore, which have reduced silica contents, are relatively fluid and flow fairly easily. Flow speeds of up to 30 mph (50 kph) have been recorded for such lavas, given suitable gradients and continual lava replenishment at source. Therefore, the fluid basic lavas tend to spread out and form extensive horizontal sheets. The flows themselves have variable surface configurations. Some are fairly smooth, some have pleated, folded or ropy looking surfaces (**ropy** or **pahoehoe lavas**), and others have blocky, rubbly surfaces (**blocky** or **aa lavas**).

Lavas with increased silica contents flow only with the greatest difficulty, the flows advancing slowly behind a steep front of avalanching debris; they form short, thick lava flows and domes. Water and gases

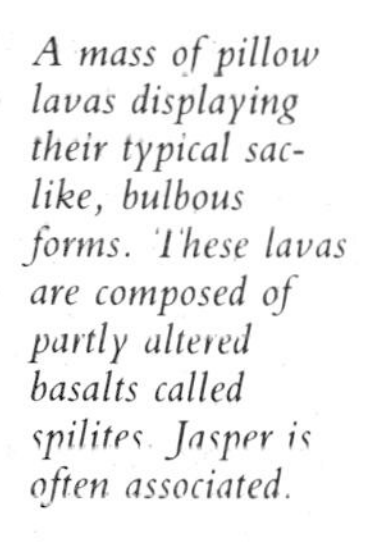

A mass of pillow lavas displaying their typical sac-like, bulbous forms. These lavas are composed of partly altered basalts called spilites. Jasper is often associated.

Ropy or pahoehoe lava, showing its distinctive folded, puckered, pleated, rope-like surface. This is caused by the outer skin of the lava cooling faster than the mobile interior. Ropy lavas are commonly composed of basaltic rocks.

dissolved in magma tend to reduce the viscosity, and consequently offset, to a degree, the effect of the silica content and thereby enhance the ability of a magma to flow.

In some cases, lavas may erupt entirely under water. They are usually basic in composition. Obviously, the extreme cooling produced under water greatly increases their viscosity, and the flow extends forward in small lobes which bud off small, sack-like lava bodies. These build up during the course of an eruption, to produce a distinctive globular deposit which is known as **pillow lava**.

In solidified basic lavas, it is quite common to find numerous gas bubbles or **vesicles**; the lavas are then described as **vesicular**. Vesicles are less common in intermediate and acidic lavas, while acidic lavas sometimes show flow banding because of the development and streaking out of compositional differences during the period of slow flow.

Intrusive igneous rocks

Intrusive and extrusive rocks are often closely associated, although some intrusive rocks represent magma which obviously crystallized entirely at depth. On land, intrusive rocks are common in areas where mountains have been formed, but little is known about igneous intrusions below the oceans. The sizes, shapes, and depths of the intrusive igneous masses and their relationships with the surrounding rocks (or **country rocks**) enable them to be classified and named. In terms of scale, intrusions are split up into **major intrusions** and **minor** or **hypabyssal intrusions**.

Intrusions may also cut across the bedding of the country rocks which are generally sedimentary, and they are then referred to as **discordant**. Intrusive rocks which have been intruded between and along the bedding planes of the country rocks are **concordant**. Major intrusions with a surface area greater than 100 square kilometres (30 square miles), steep contacts with the country rocks, and no observable base are termed **batholiths**. Projections from the upper part of such masses are **stocks** and **bosses**. Stocks are smaller but similar in shape to batholiths, while bosses are smaller again and possess circular cross-sections. Rocks which compose these discordant intrusions are referred to as **plutonic**, and they are often acidic. The intrusion of such bodies is attributed to the process

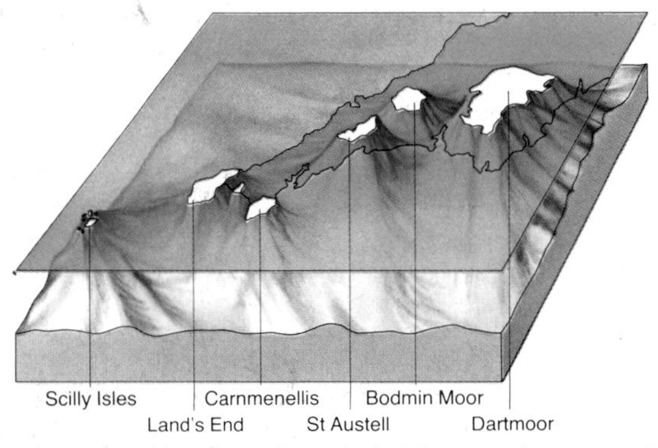

Resistant granite stocks, rising from a larger batholith, create the scenery of S.W. England. Overlying country rocks have been eroded away to reveal the stocks.

called **stoping**. Here, rising magma is squeezed into and along cracks and fractures in the country rock, causing blocks to become detached and to sink into the magma, thus permitting the magma to continue its upward advance. Occasionally, large, thick, roughly concordant sheets of basic intrusive rocks may be found. Surface areas approach, in some cases, 6000 square kilometres (2300 square miles), while thicknesses can be up to 6 kilometres (4 miles). Layering (*see* gabbros page 98) is frequently developed within such bodies.

Minor or hypabyssal intrusions tend to be rather more varied in shape and can be either concordant or discordant. These intrusions are all thought to originate from a large magma body, or **magma chamber**, at great depth. **Dykes** are vertical or oblique discordant intrusions, varying in size from a few centimetres to many metres across, but generally they are about 3 metres (10 feet) wide. When exposed at the surface, characteristically they run in almost straight lines and may extend for a short distance or for many kilometres. A common intrusive igneous rock in dykes is dolerite (*see* page 102). If the dyke rock is harder than the surrounding country rock into which it was intruded, it will form a linear upstanding feature in the landscape, following a long period of erosion. Conversely, if the dyke rock is softer, then a linear depression will be produced following erosion. Dykes occasionally occur in parallel or radial groups, or **dyke swarms.**

Sills are sheet-like, concordant intrusions, and were formed in situations where the magma found it easier to exploit subhorizontal planes of weakness (usually along bedding planes) rather than rise higher. The magmas need to be rather fluid to produce this sheet-like form, so sill rocks are usually basic in composition. Sometimes, along the length of a sill, the injected magma changes from one level to another, following planes of weakness, and it is then said to be a **transgressive sill**.

In some cases, the component layers of country rock may be uplifted into gentle folds by the pressure of the intrusive magma. A dome-like mass of igneous rock with a more-or-less flat floor is then produced. Such a 'mushroom-shaped' body is called a **laccolith**, and may be fed by either a lateral or a vertical pipe.

Sometimes, magma may be intruded into country rocks which had previously been folded. The floor and the roof of the intrusion then have the same shape, and they are usually fed by vertical pipes. Such intrusions may be called **lopoliths**, which are saucer shaped or concave upwards, or **phacoliths**, which are convex upwards. Some of the large, sheet-like, basic intrusions develop a lopolith form. Due to their large size, perhaps they should be called 'megalopoliths'.

Certain minor intrusions, when exposed, are found to have a ring-like, circular shape. These are the **cone sheets** and **ring dykes**. Cone sheets are dykes which form part of a conical surface and slant inwards to a common focus, usually a magma chamber. The almost cylindrical ring dykes are often intruded around the margin of an area which has subsided. Subsidence is thought to be related to decreased pressure in the underlying magma chamber.

Texture of igneous rocks

The texture of any rock is the relationship between the crystals or grains comprising that rock. A great deal can be learned about the history of a rock by examining its texture. To study a rock's texture in detail, it should be examined under a microscope, but you can still learn a good deal about it using a hand-lens or, with coarse-grained rocks, simply with the naked eye. Igneous rocks are composed of discrete crystals so that they are referred to as **crystalline**. These crystals closely interlock, with no space between them. The crystals are also randomly distributed – a diagnostic feature of igneous rocks. When you look at an igneous rock try, in the first place, to identify any crystals with the naked eye. Coarse-grained igneous rocks have crystals with an average size greater than 5 mm (0.2 in); medium-grained between 1 and 5 mm (0.04 to 0.2 in); and fine-grained with crystals smaller than 1 mm (0.04 in). Finally, try to estimate whether the crystals making up the rock are all roughly the same size (**equigranular**) or whether a whole range of different sizes are present (**inequigranular**). In some cases, large visible crystals, called **phenocrysts**, are surrounded by a groundmass of very much smaller crystals. This texture is termed **porphyritic**, and the rock is known as a **porphyry**.

Igneous textures are controlled by how quickly or slowly a magma or lava cools, the order in which the minerals crystallized, viscosity, and any movement the cooling body may have undergone. Regarding cooling, a useful rule of thumb is that the faster a magma or lava cools, the finer the final grain size will be. The rate of cooling depends on the size and location of the magma body. Large bodies cool more slowly than small ones, and lavas cool much more quickly than deep intrusions which are usually surrounded by country rocks that are themselves slightly warm. Extremely rapid cooling produces glass, because the atoms have not had time to form themselves into the regular atomic structures of which the silicate minerals are composed. As glass does not possess crystals, it is not crystalline.

Rocks from major intrusions are usually coarse grained. Rocks from minor intrusions, or hypabyssal rocks which, being nearer the surface, cooled rather more quickly, are usually medium-grained or porphyritic. Extrusive rocks cool very quickly, and are usually fine-grained, porphyritic, or even glassy. Slow cooling is also encouraged by low viscosity as well as by the presence of dissolved water.

Crystallization occurs over a range of falling temperature, so that certain minerals, with higher 'freezing points', form before others. Early formed crystals grow freely in the slowly cooling magma, and assume their characteristic or **euhedral** crystal shape. With continued cooling,

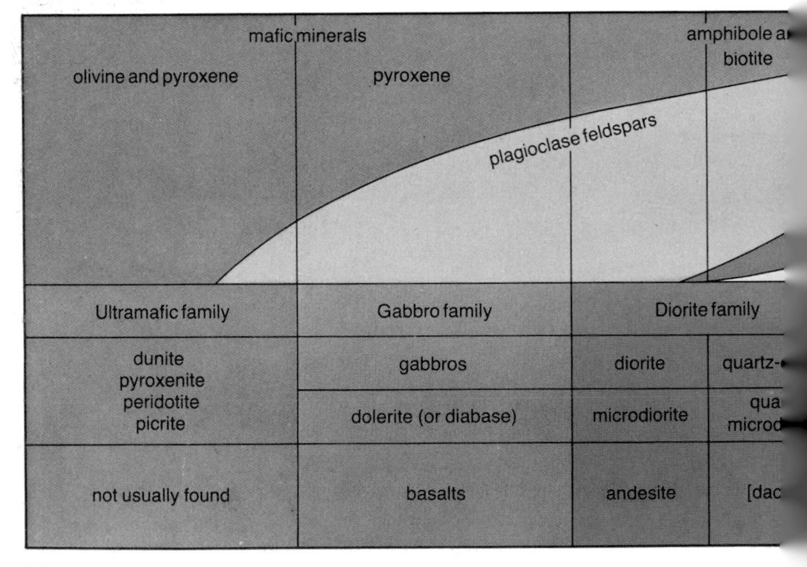

more crystals form, and may interfere with one another's growth to some extent. Characteristic crystal shapes are partially modified to become **subhedral**. When many crystals form simultaneously, unrestricted growth is severely inhibited and compromise or irregular crystal shapes are developed; such crystals are described as **anhedral**.

Many magmas start to crystallize at depth, and it is here that the phenocrysts of a porphyritic texture form, growing freely in the magma. If this crystal-liquid mixture is erupted on to the surface, the remaining liquid crystallizes rapidly to form a very fine groundmass surrounding the much larger phenocrysts. As with all igneous rocks the mineral content reflects the composition of the original magma, whereas the texture reflects not only the size and shape of the crystals, but also the cooling history of the rock.

Classification of igneous rocks

Many classification schemes have been proposed, but no one scheme will ever be ideal because igneous rocks grade continuously between one another. The best scheme is one that is easy to use in the field and in the laboratory, and emphasizes the relationship between similar igneous rocks. The system that satisfies these requirements is one that combines the mineral composition of the rocks and their texture. You should be able to work out, for example, that a rock containing roughly 30 per cent biotite, 40 per cent plagioclase feldspar, 20 per cent alkali feldspar, 10 per cent quartz, and with a coarse grain size, would be a granodiorite. Remember, though, that only the more common 'families' of igneous rocks are shown here.

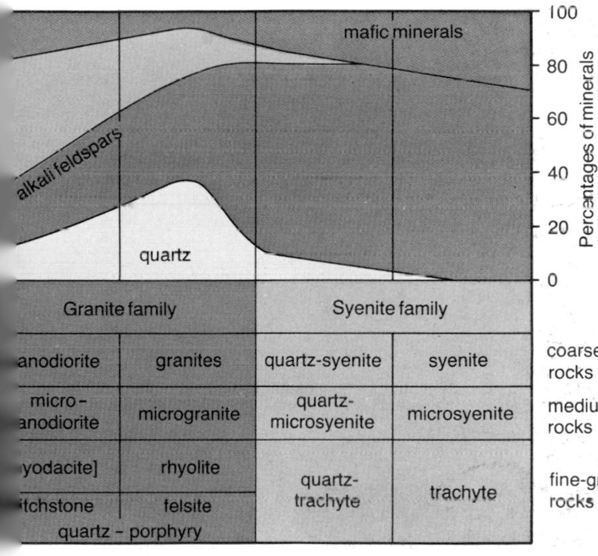

A chart depicting the classification of common igneous rocks, based on the percentage of silicate minerals present, and the average grain size. Names in square brackets represent igneous rocks which are either rare or not described here.

Sedimentary rocks

Only 5 per cent of the Earth's crust consists of sedimentary rocks, yet 75 per cent of the rocks exposed at the Earth's surface are sedimentary. Coal is a sedimentary rock, and these rocks contain useful minerals as well as providing reservoirs for water, oil, and gas, so they are important economically. Their formation is of interest, too, not least because many of the processes by which they are laid down can be observed at the Earth's surface today. The processes by which sedimentary rocks are formed are best shown in the diagram below which expresses the way in which the major rock types are linked.

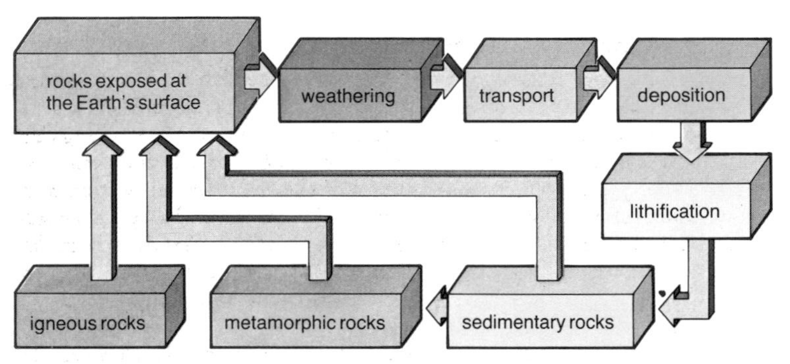

Flow diagram depicting the sedimentary rock cycle. During and after lithification the rocks may be buried to great depths before finally being exposed.

Weathering

Rocks at the Earth's surface are subject to weathering by chemical action of acid and alkali salts in rain and ground water. They are also broken down mechanically by plant roots, expanding ice, and alternate heating and cooling. In a moist climate, plant roots break up the surface of an exposed rock allowing water to penetrate. This water assists chemical breakdown and, once inside the cracks and crevices of the rock, it can freeze. As it freezes, the expanding force of water is so great that the rock can shatter. Frost shattering is a major process of mechanical breakdown and is responsible for the accumulations of angular rock fragments, known as **scree** deposits, at the foot of steep mountain slopes. In hot deserts, rocks can fracture as they expand during the heat of the day and contract during the cold night. Tunnel walls in a mine can collapse with explosive force because of the pressure of the surrounding rock. This happens to some extent at the Earth's surface. Many rocks that are now exposed were once buried under great masses of overlying layers of rock. Once the overlying rocks have been removed by erosion, the release of pressure can cause fractures parallel to the rock surface (**sheet joints**).

Alternate freezing and thawing of moisture within cracks shatters the rock surface and the fragments accumulate at the foot of the mountain slope forming a scree with a slope of 25°–35°. A cemented scree is called a breccia.

Transport and deposition

Once it has been weathered, rock material may be removed by transport. Some minerals are removed in solution and some remain undissolved. A stream can transport undissolved particles by bouncing them along the stream bed (**saltation**); if the particles are small enough, they may be carried in **suspension**. An alpine stream in spring is often opaque and milky with suspended sediment. Occasionally, very fine dust gets carried high into the atmosphere and is carried great distances. Red dust from the Sahara sometimes travels as far as Britain and northern Europe to coat windows and cars. Wind can also transport particles by saltation along the ground and in suspension as dust.

The weathered rock can be deposited as a sediment by three processes:

1. **Settling out of particles**. Fragments of rock carried in suspension drop out of the transporting medium (wind or water) when the velocity decreases. The size of particle that settles is controlled by this velocity. This process can lead to **graded bedding** as in greywackes (*see* page 127) where the largest-sized grains settle out first and are followed by increasingly fine grains. This can be used to indicate which way up a tilted bed was originally.
2. **Chemical precipitation**. Minerals, such as calcite and dolomite, may occur as precipitates from fluids in rock pores and also directly from sea water (*see* 'micrite', page 134 and 'evaporites' page 142). The 'fur' which you can find in kettles in which hard water has been boiled is a good example of chemical precipitation. In much the same way, if sea water, in which the circulation is restricted, is heated, some of the chemicals that were in the solution may precipitate out.
3. **Organic precipitation**. All marine creatures which have hard parts, such as skeletons or shells, obtain the materials from which they are made either from sea water or from eating other organisms. These creatures eventually die and their remains become part of the sediment on the sea floor. Some organisms, such as corals, often build a rigid framework which may be preserved relatively intact upon burial.

The final sediment may consist of one or more of these processes.

Lithification

Sediments are converted to sedimentary rocks by the process of lithification. The process known as **diagenesis** includes all those changes that take place in a sediment near the Earth's surface at low temperature and without deep burial, and gives rise to lithification. As a sediment is gradually buried by material that accumulates at the surface, water is expelled and the sedimentary particles are compacted. This compaction can realign flat and elongated grains and even weld particles together.

Cementation results from the crystallization of minerals dissolved in water. This cement is precipitated in the spaces or pores between particles. If the cement does not totally block the points at which the pores interconnect, water can percolate through the network of pores. This type of rock is described as **permeable** and, if it contains enough water to be recoverable, it is called an **aquifer**.

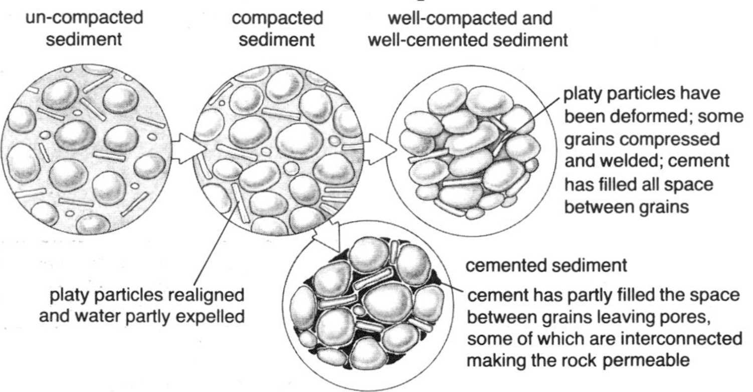

Above: Schematic diagram showing the effects of compaction and cementation upon a sediment. If the sediment is cemented at an early stage, the cement can form a rigid framework and prevent further compaction during burial.

Left: London Clay cliffs on the Isle of Sheppey in England. Fossils are readily found in some clay. Often they are made of pyrite but they may also be made of mud, and disintegrate when washed. Clay can be treacherous to build upon; cliffs such as these often collapse.

Classification of sedimentary rocks

Rocks may be named after the places where they were first recognized, such as Green River Shales of Wyoming, USA or Kimmeridge Clay from Dorset, England. They may also be named on the basis of texture or colour, such as tilestone and black shale. Rocks can even be named for the use to which they may be put, for example, the Millstone Grit which was used in the past for millstones. Classification may also be based on origin, for instance, a fault breccia. In this book sedimentary rocks are discussed in terms of the two main groups generally accepted by geologists today: **clastic rocks** which incorporate mostly transported particles, and **non-clastic rocks** which are chemically and biologically precipitated.

Clastic rocks are subdivided according to grain size into three main groups: **mudrocks**, **sandstones** and **conglomerates**. These are further subdivided on the basis of their composition and texture. Non-clastic rocks are subdivided according to their chemical composition, structure and origin.

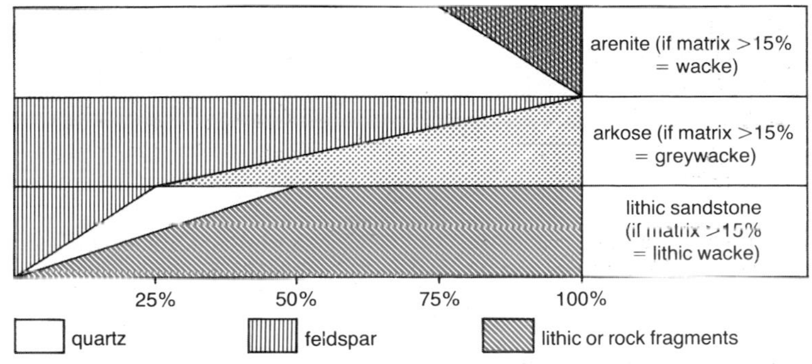

A subdivision for sandstones which have grains visible to the naked eye but less than 2 mm (0.08 in). Areas of overlap denote mixtures of components.

TEXTURE AND TEXTURAL MATURITY **Grain size** is a particularly important distinguishing characteristic for clastic rocks, and the classifications are based mostly upon this feature. If you look at a clastic sedimentary rock using a hand-lens, you can also estimate the degree of sorting. During transport and sorting, grains will batter against each other and become rounded. When the grains are coarse and angular the rock is termed a grit.

The relationships between grains and the matrix are referred to as grain fabric. Some grain fabrics which are used in classification are shown in the illustration. A combination of different textures, grain fabrics and shapes give rise to varying degrees of textural maturity. A rock becomes mineralogically mature as weathering and other sedimentary processes break down individual minerals and eventually produce sediments composed almost entirely of one mineral type. This sediment, produced by the breakdown of rocks, will initially be poorly sorted and the constituent grains will be very angular. If the material is transported over a long period of time, it will become well sorted and the grains will be well rounded, in which case it is said to be texturally mature. Try examining local sand, even builders' sand, to assess its textural maturity.

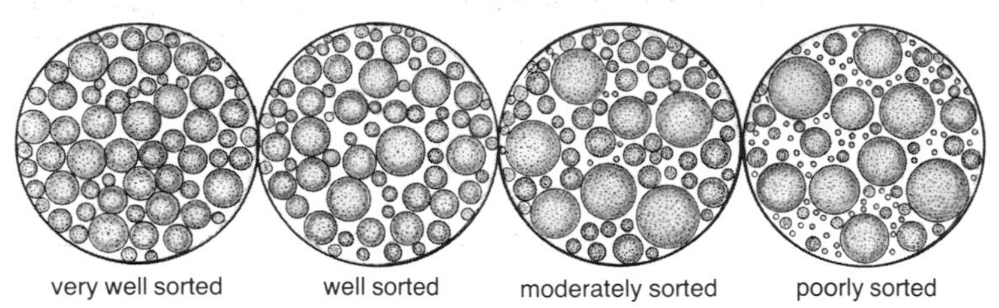

Above: Charts for estimating sorting visually. Using a hand-lens, compare your sample with each chart and estimate the closest comparison.

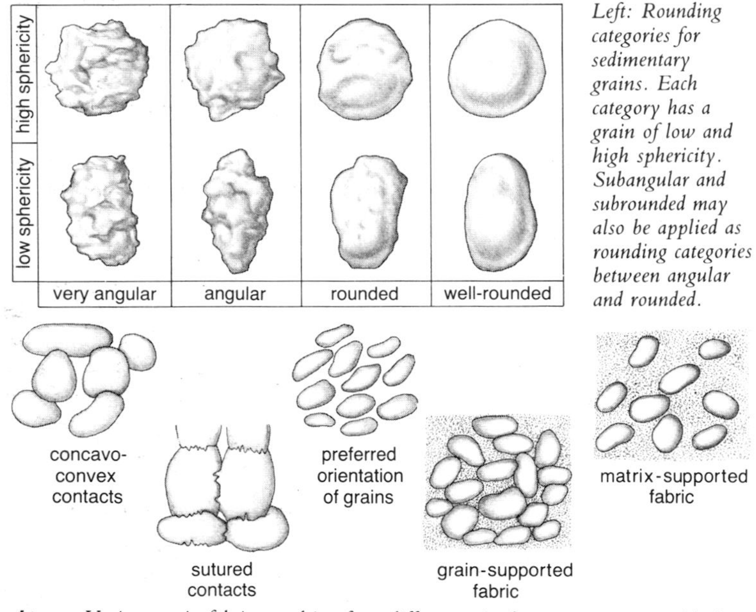

Left: Rounding categories for sedimentary grains. Each category has a grain of low and high sphericity. Subangular and subrounded may also be applied as rounding categories between angular and rounded.

Above: Various grain fabrics resulting from differences in the processes responsible for forming the sediment, and also in degrees of subsequent compaction.

Sedimentary structures

When sediments are laid down, structures are formed which stem from the behaviour of the individual grains and the way in which they are transported.

BEDDING PLANES represent surfaces of deposition. In shales (*see* page 118), the rock splits along bedding planes where platy minerals are roughly parallel to the planes. Often bedding is not defined by a recognizable surface but is only detectable by changes in colour and/or composition.

CROSS-BEDDING OR CROSS-STRATIFICATION is also occasionally known as **false bedding** and is formed by the migration of ripples as grains move up the shallow slope of the ripple and into the trough. The trough is maintained during the migration by eddies. Ripple migration can be clearly observed where the waves lap on the beach nearest the shore itself. If you cut a small trench in this sand you may see the cross-bedding which often has symmetrical ripples because the current flows both up and down the beach. Ripples forming in rivers and streams will be characteristically asymmetrical with the steep lee side facing the direction of flow.

Cross-bedding can take several forms. **Tabular cross-bedding** is formed by the migration of relatively straight ripples while **trough cross-bedding** is caused by crescent-like, scoop-shaped ripples. Cross-bedding can indicate which way up a bed of rock was originally laid down because the migration of ripples can only truncate the tops of the ripples below them. Where mud is deposited alternately with ripple migration, often on tidal flats, **flazer bedding** and **lenticular bedding** can occur.

The Entrada Formation Church Rock of New Mexico in the United States is a Jurassic sandstone with cross-bedding formed as a result of wind-blown sand in dunes.

Shown here is a sequence of bedded sandstones and clay-rich beds. An outcrop such as this should have a bedding plane exposed where you can measure the dip and strike.

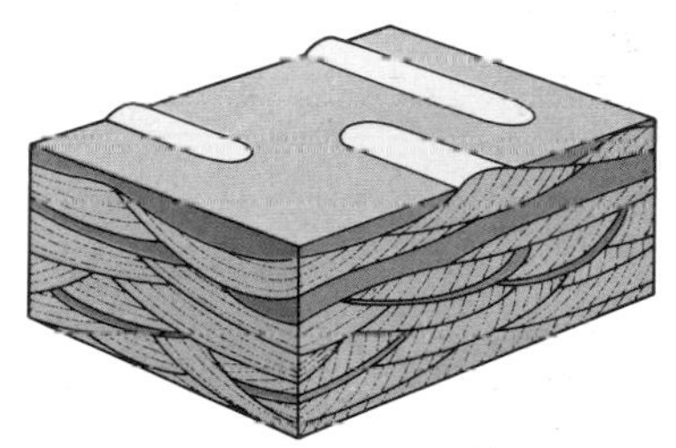

Flazer bedding; sorting of mud and sand.

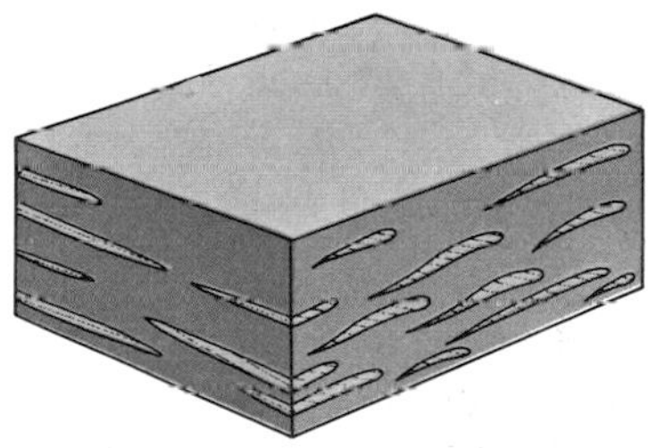

Lenticular bedding caused by migrating ripples.

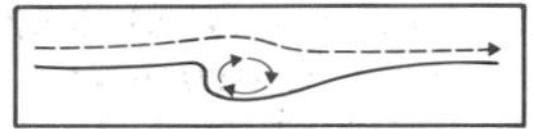

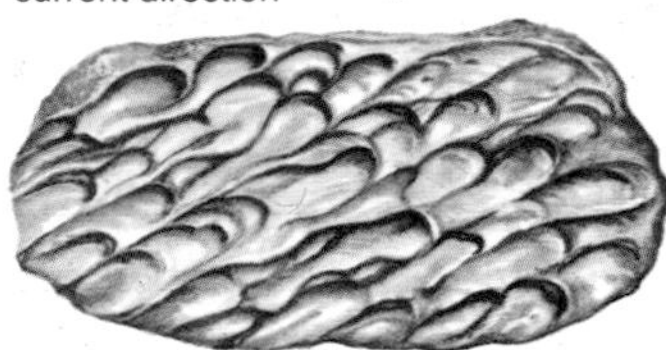

Above: Flute cast on a bedding surface with cross-section (top).

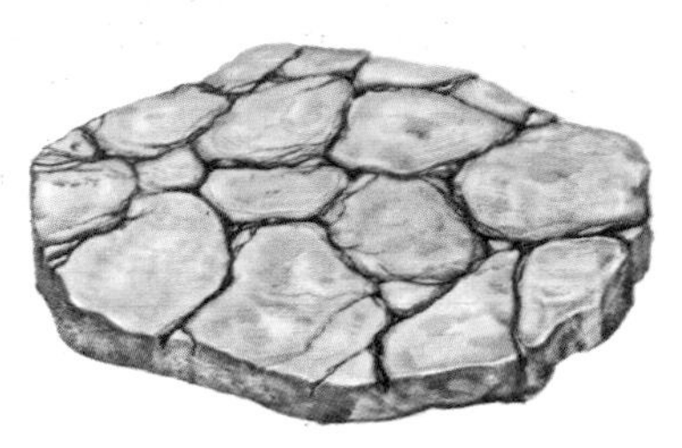

Above: Shrinkage cracks in drying mud.

Right: channels form particularly in fluvial environments. Erosion 'scours' the sediment surface forming channels which are later filled in by sediment.

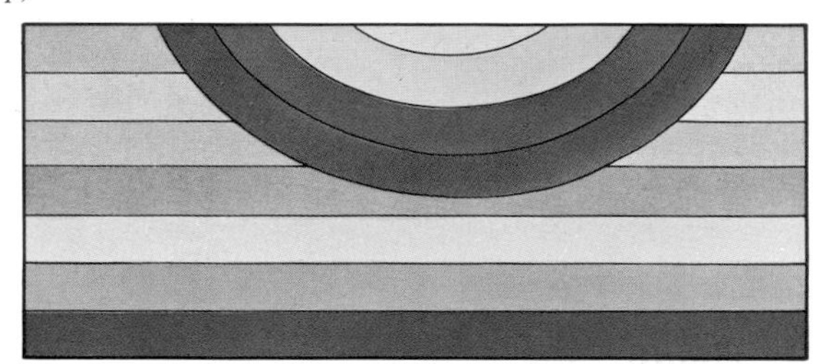

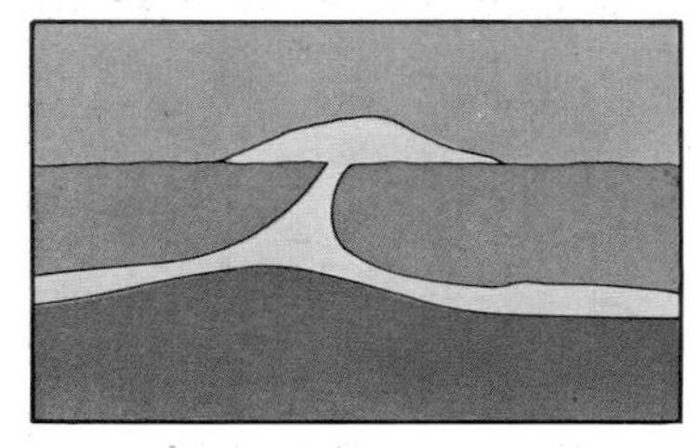

Above: During burial some clay-rich sediments remain fairly fluid. Load casts and flame structures (right) may form, and can escape upward as mud volcanoes (left).

FLUTE MARKS are heel-shaped depressions which are bulbous at the upstream end and widen downstream to merge into the bedding plane. The diagram shows how a flute mark is eroded in a cohesive mud surface by sand-laden currents which fill the depression with coarse sediment. These are common in **turbidites** (*see* page 37).

TOOL MARKS are formed when grains or 'tools' are dragged along the bedding plane leaving a groove which is roughly parallel to the direction of the current flow. When the current speed is great enough, these tools bounce or **saltate** along the bed and leave **saltation marks**, sometimes termed **prod marks**.

CHANNELS AND SCOURS are erosional features. Channels may be metres or more in size while scours are smaller and occur within beds or at their base. Short-lived erosive currents cause oval to elongate hollows in the bedding plane; these scours later become filled with

sediment. Longer-lived erosive currents cut channels, which tend to be more laterally persistent, and form larger-scale features. Channels were cut in the Grindslow Shales in Derbyshire, England, and these channels were filled by sandstone; their bases often show **groove marks**.
SHRINKAGE CRACKS occur in many fine-grained rocks and may be seen today in any mud as it dries out in the sun. The cracks join to form polygonal shapes and sediment can infill the cracks, thus preserving them.
RAINPRINTS are often found associated with shrinkage cracks. The raindrops form small rimmed depressions through impact on the soft, exposed surface of fine-grained sediments.
TRACE FOSSILS are the preserved tracks, trails, and homes of animals. They may be branching burrows, worm-like grazing traces, or even vertebrate footprints. If the sediment is intensively disrupted by the activity of organisms (**bioturbated**), sedimentary structures and even some grains may be destroyed.
LOADCASTS form when a bed of sand is deposited upon fine mud. The sand tends to press downwards and form bulbous projections into the mud. This process may go further if the mud is waterlogged, and **flame structures** or even **mud volcanoes** may form.

Unconformities

Unconformities are breaks in the succession of sedimentary rocks following a period of erosion and/or a period when sediments are not being deposited. If the rocks below the break are tilted or folded there will be a difference in the angle of dip between the rocks on either side of the unconformity. In such a case the break is called an **angular unconformity**.

Karst surfaces are unconformities which occur where limestones have been dissolved by rain water. The surface becomes irregular, with potholes and caverns which may be observed today in areas such as the Peak District in Derbyshire, England. These surfaces may be preserved geologically (**palaeokarst surfaces**) and may have thin fossil soil layers with root tubes above them.

When rocks are exposed, eroded and then buried by later sediments the previous erosion surface is called an unconformity. If the underlying beds have been tilted before sedimentation recommences the unconformity is called angular, as shown here.

Sedimentary environments

Certain sequences of sedimentary rock types and their associated structures can indicate the environment in which the rocks were laid down. These sequences are known as **facies**.

Facies are groups of sedimentary rocks with features that distinguish them from other groups. They can be described in terms of the way in which they were deposited, such as a river delta facies, the environment, such as a beach facies, or in terms of the sediment itself, such as a cross-bedded sandstone facies. A facies refers to the sum total of all the features of a sedimentary rock group including its fossils and sediments. Facies may cover relatively small areas but they can also cover areas of hundreds of square kilometres, such as for example, the Yoredale deltaic sequence of northern England or the Mesozoic desert sandstones of the Colorado Plateau, USA. The distribution of geological facies can be used to reconstruct past geographies.

Vertical changes from one facies to another may be caused by depositional processes alone, such as the change in sedimentation which occurs between a river and its mouth or delta, or by changes in sea level.

Sea-level changes occur especially during ice ages when huge quantities of water are locked up in the ice caps. During past ice advances, for example, the continental shelf around Europe was exposed. When the ice melted, the sea level rose and engulfed the shelf. This kind of sea-level change occurs worldwide and is called **eustatic**. After the ice retreats, local, or **isostatic**, changes in sea level may be produced by movements of the land or the sea floor. The weight of the ice depresses the land surface and, when this weight is removed, the land can slowly rise again. Changes can also occur as a result of movements such as those caused by sediment compaction or **faulting** (*see* page 41).

Sandstone environments

The following environments are not exclusively those of sandstone rocks although, in modern examples, sand and silt are the dominant sediment types. It might be more precise to refer to these as **clastic environments**.

DESERT ENVIRONMENTS Today, these are generally found in subtropical areas. In both modern and geological sequences areas of wind-blown (aeolian) sand with large-scale dune cross-bedding are common. Various types of dunes occur, such as linear or seif dunes and crescent-shaped or barchan dunes. Water still plays its part in these environments, giving rise to temporary stream channels and alluvial fans. The fans occur where flash floods laden with sediment discharge from mountainous areas and dump it at the mountain foot. This water will flow out on to the desert plain and form a temporary **playa** lake until it dries up or soaks into the ground. Wind-blown desert sandstones, often quartz arenites (*see* page 124), are texturally mature and may be stained red with iron.

A small barchan dune in southern Tunisia. These crescent-shaped dunes are constantly moving, with the horns of the crescent pointing downwind. A cross-section through the dune would show cross-bedding.

RIVER ENVIRONMENTS These are a complex combination of sedimentary environments. The sediments are deposited by rivers and streams as alluvial fans, various braided networks, and meandering rivers. The sedimentary rocks deposited range from the finest mudrock to the coarsest conglomerate (*see* page 128). The sequences that occur and the environments responsible for each rock type are summarized in the diagrams.

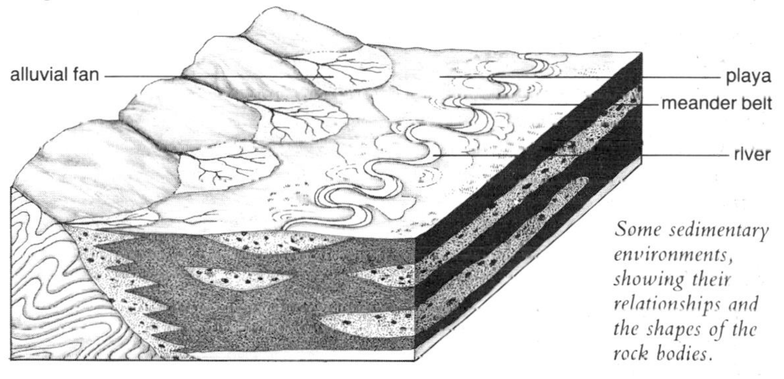

Some sedimentary environments, showing their relationships and the shapes of the rock bodies.

DELTAIC ENVIRONMENTS These are found where a river flows into an open body of water such as a lake or the sea. The current speed rapidly falls and the sediment being carried is dropped; the coarser grains are dropped nearest the river mouth and the finer sediments further out. Much clay material is deposited owing to **flocculation**, where fine particles stick together and behave like coarser ones. Lake deltas are similar to marine deltas but generally they are smaller. Two common types of delta are the birdfoot delta, such as the modern Mississippi, and the lobate delta such as the Nile delta. In a simple delta, three main types of bedding may be identified which can be seen on a smaller scale in cross-bedding.

1 bottomset beds, comprising the finer grains and deposited on the sea bed at quite a distance from the river mouth;

2 foreset beds build outward over the bottomset beds and form the slope of the delta front;
3 topset beds are deposited on top of the foresets and are composed of coarser sand grains, being closer to the river mouth.

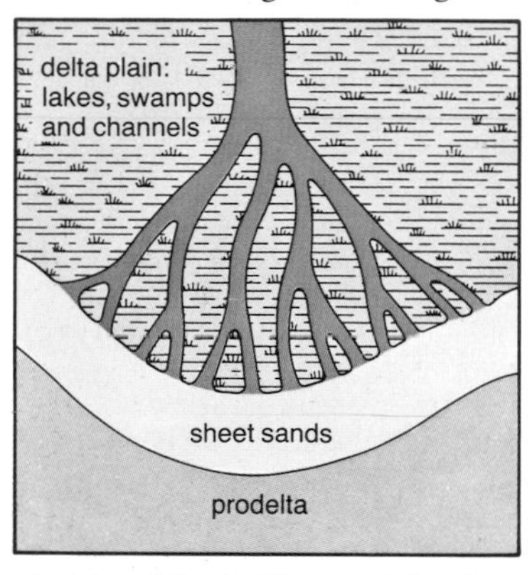

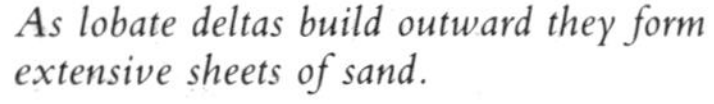
As lobate deltas build outward they form extensive sheets of sand.

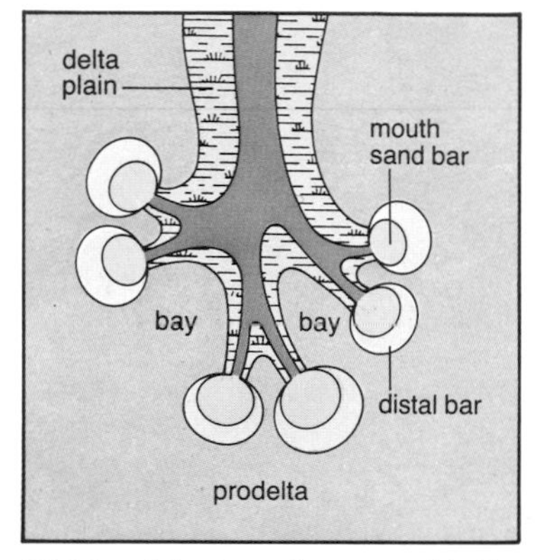

Birdsfoot deltas form elongate sand bodies which often overlap earlier deltas.

BEACHES AND SHORELINES Here, there is a variety of different sedimentary environments, and many of them can be seen at the present. Beaches are linear bands of sand or pebbles which form a shoreline. Offshore bars are banks of clastic material separated from the land by a lagoon. Sometimes bars can build out from a beach to form spits. The south coast of England has several examples of bars, lagoons, and spits, notably Chesil beach in Dorset (spit), and Slapton Ley in South Devon (lagoon and spit, connected to land at either end).
CONTINENTAL MARGINS AND DEEPWATER BASINS These are dominated by sediments which slide and slump down the continental slope. Slumps and slides can involve differently

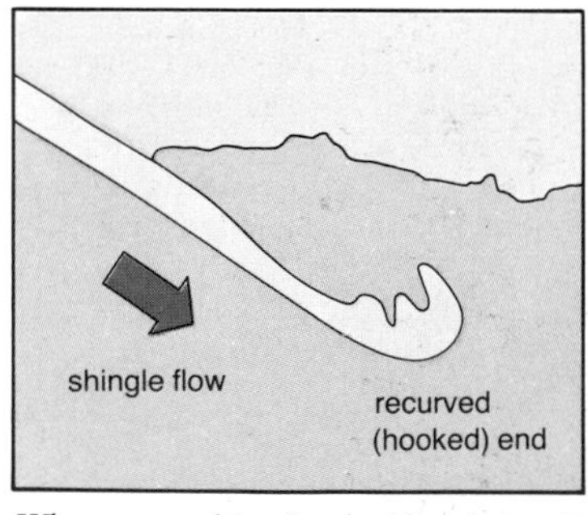

Where waves hit a beach obliquely a spit can build out from a headland.

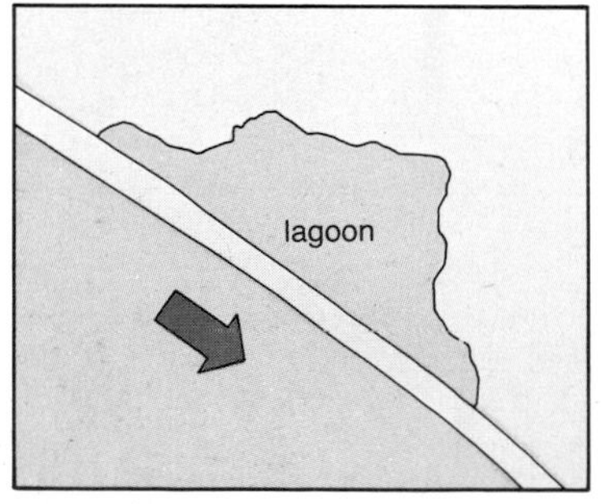

If a spit forms across a bay, joining the opposite shore, a bar is formed.

sized masses of sediment, often with folding and brecciation (*see* page 128). When downslope currents keep sediment in suspension, turbidites may be deposited when the current speed falls. The deposits are characterized by sedimentary structures such as graded bedding and flute and tool marks.

Limestone environments

Most of the environments which produce sandstones also give rise to limestones, although certain environments yield distinctive types of limestone.

SUBAERIAL ENVIRONMENTS These are areas of existing limestone rocks which are exposed at the Earth's surface and which may go into solution. Lime-rich soils form distinctive vertically zoned, often nodular deposits which usually have well-defined tubes. Soils of this kind may be found in Tarragona and elsewhere in Spain and in the Mediterranean region; they may also be found in the Guadalupe Mountains in New Mexico.

Travertine, tufa, or **sinter**, is an accumulation of calcium carbonate which forms in hot springs and in karst areas. It may often be seen in limestone caves as stalactites hanging from the roof. Mammoth Hot Spring in Yellowstone, USA, has abundant travertine accumulations mainly due to the activity of algae.

LAKE ENVIRONMENTS These may produce limestones which are often similar to marine limestones, such as algal stromatolite (*see* page 149) and ooid shoals (*see* below). A characteristic feature of many lake limestones is the layering, called **varving**, which forms as a result of seasonal variation in deposition. The Green River Formation in Wyoming and Colorado in the USA shows varving.

TIDAL FLATS Vast areas which may only be covered by sea water at high tide or possibly only at exceptionally high tides are called tidal flats. Limestones of tidal flats are mostly micrites (*see* page 134) and are often disturbed by the activity of living creatures. Shrinkage cracks, rainprints, and other sedimentary structures are common in this environment. **Sabkhas** are tidal flats where evaporites (*see* page 142) form.

Evaporites can also form inland. This inland sabka shows laminated gypsum at Chott Djerid in Tunisia.

Inland sabkhas can also occur. **Supratidal** areas, or those which are very rarely flooded, are also included in this category.

LAGOONS AND RESTRICTED BAYS These are located behind barriers such as reefs or carbonate sand shoals. Green algae are common and the sediments are often composed almost entirely of faecal pellets. Lagoonal limestones, such as the Lowville limestone in New York, USA, are common in the geological record.

SHELF ENVIRONMENTS These actually constitute a range of environments including lagoons and shoals (*see* below). The various facies, however, form a series of associations, many of which can be observed at the present day in areas such as the Bahamas or the Arabian Gulf. Shelves which show gradational facies changes between broad facies zones, like that of the Arabian Gulf, tend to be gently inclined **ramps**. The **drop-off** shelf types, such as the Bahamas, tend to have clearly defined linear facies belts parallel to the shelf edge which drops off rapidly into a deep basin. A combination of a falling sea level together with compaction and lowering of a shelf sediment can result in a series of **shoaling-upward** cycles. An entire cycle might consist of deeper-water micritic limestones with overlying shoals, tidal flat deposits, and often supratidal evaporites at the top of the sequence. These cycles are often incomplete and may occur repeatedly.

SHOALS These are shallow-water, carbonate sand deposits and may form as spits building out from a beach or as bars (*see* page 36). They form where wave energy stirs up the sea bed, and are often composed of shell fragments, ooids, and peloids with very little micrite due to 'winnowing' by the waves. Frequently, these carbonate sand bodies display cross-bedding and ripples.

REEFS AND CARBONATE BUILD-UPS Generally, these are biologically constructed features which form high points on the sea bed that may be exposed at low tide. Carbonate build-up is a term now widely used to describe limestone accumulations where there was an original topographic feature. Often these are important reservoirs for oil and gas. Many reefs, particularly those at shelf margins, may be divided into three zones:

1 **back-reef** which consists of reef debris near the reef itself and passes shoreward into a lagoon with flat-bedded sediments.
2 **reef flat** which is covered by about 1 to 2 m (3 to 6 ft) of water and experiences abundant growth of the reef-forming organisms. The seaward margin of the reef front often has a series of surge channels with intervening highs, known as **spur and groove morphology**. The reef itself is often massive with no layering to be seen in outcrop.
3 **reef front** which is a steep, often vertical, slope which passes down into a **talus slope** of debris derived from the reef itself.

Ancient examples of reef zones are common, for example, Devonian reefs at Windjana Gorge, Napier Range in Western Australia and the Upper Triassic Steinplatte Reef in the Tyrol, Austria. Some ancient carbonate build-ups comprise mostly micrite and are probably the result of this being trapped by plants that have not been preserved.

Evaporite environments

Geologically, evaporites are restricted to fairly arid areas. Modern evaporites are forming on the Arabian Gulf coast, along the Texas coast, and in the Coorong in Australia. The common evaporite minerals are gypsum, anhydrite, halite, and dolomite (*see* pages 142 and 143). Thick, layered sequences of evaporites require a great deal of brine evaporation and usually tend to represent environments other than coastal sabkhas (*see* page 37).

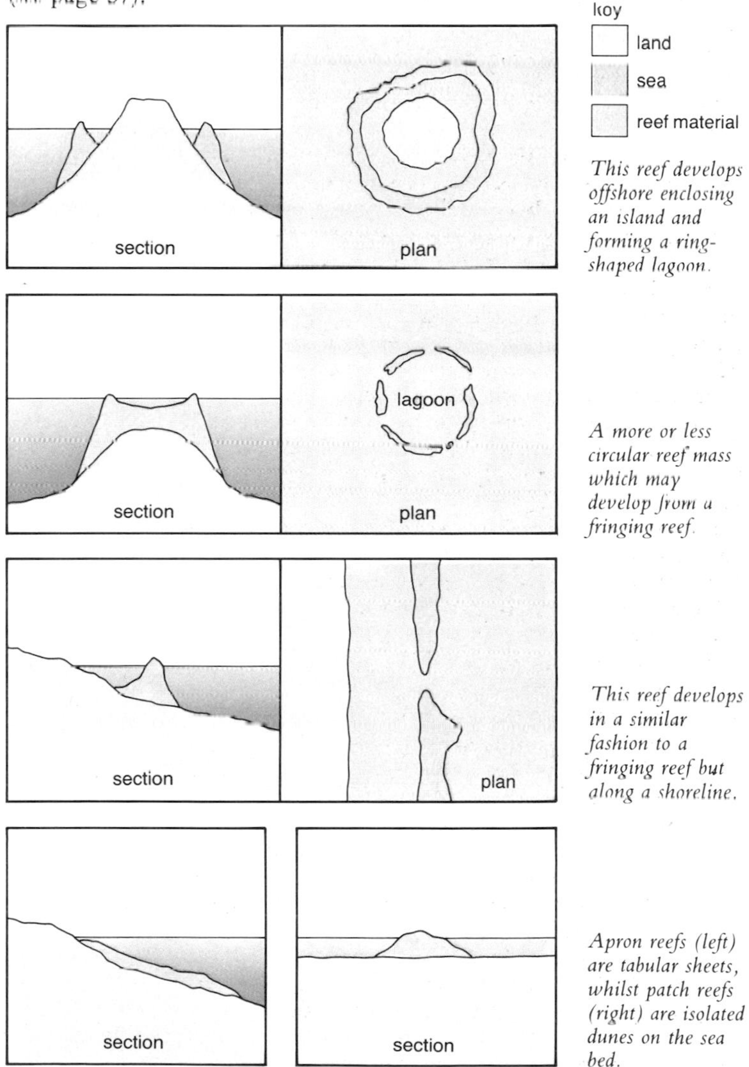

This reef develops offshore enclosing an island and forming a ring-shaped lagoon.

A more or less circular reef mass which may develop from a fringing reef.

This reef develops in a similar fashion to a fringing reef but along a shoreline.

Apron reefs (left) are tabular sheets, whilst patch reefs (right) are isolated dunes on the sea bed.

Folding, joints, and faulting

Folding

When a sedimentary rock is laid down, it forms a more-or-less horizontal bed. When it is buried, however, it is subjected to compressive forces within the Earth's crust. Because of the intense temperature and pressure, the rocks can be bent like putty into three main kinds of folds.

Anticlines are folds that have a crest and the older rocks in the centre of the fold, while **synclines** are trough shaped with younger rocks in the centre. Step-like folds are called **monoclines**. Although folds were formed deep in the crust, millions of years of erosion may have exposed them but they are still usually covered with soil and vegetation so that outcrops may be rare. The many different forms of folds give rise to characteristic outcrop patterns and can control the shape of the landscape. Sometimes beds can be completely overturned. This is not always obvious and you should always look for structures which indicate which way up the beds were originally laid down.

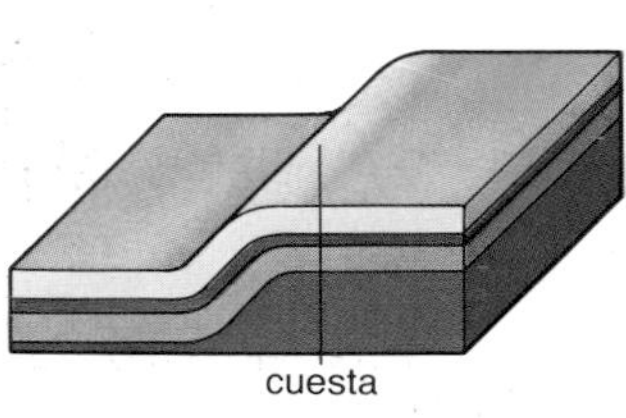

A stair, step or monocline fold.

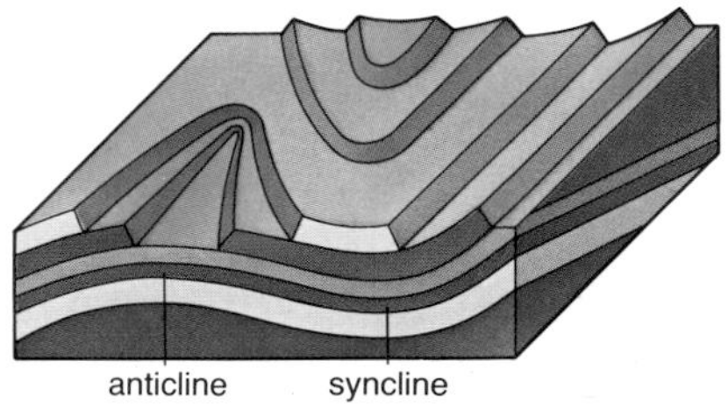

When a fold is not parallel to the Earth's surface it is described as plunging.

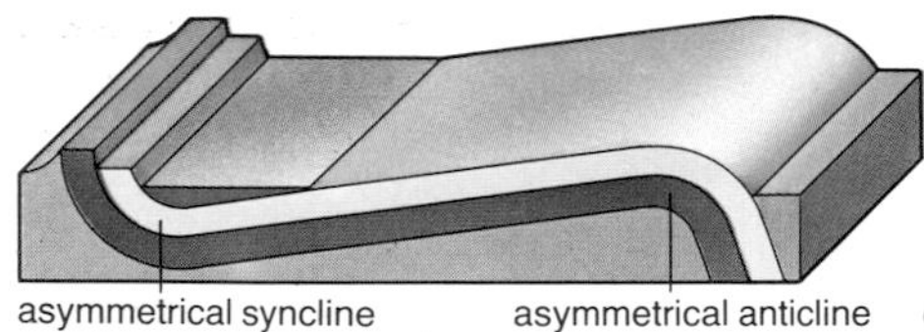

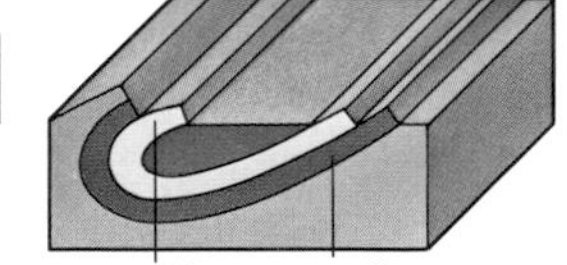

Diagrams showing some different fold styles.

Joints

These are fractures in rocks where no observable movement has occurred between the rocks on either side. Joints can often be mistaken for bedding planes and you should be careful to look for sedimentary structures, colour, and variation to identify the position of the bedding planes.

Various forms of joints can occur. **Sheet joints** are caused as a result of pressure release upon exposure. **Hexagonal joints** can form in igneous rocks as they cool (*see* page 108). Joints also form in association with folding and can help to demonstrate the presence of folding which is not obvious.

Faulting

Faults are fractures in rocks where the rocks on either side of the fault have moved in relation to one another. Faulting often occurs when rocks are not buried to depths great enough for folding to occur. The temperature and pressure are too low for the rocks to be flexible and, consequently, they are brittle and can fracture. When a fault runs approximately parallel to the strike of a bed, it is called a strike fault, and when in the direction of dip (*see* page 10) it is called a dip fault.

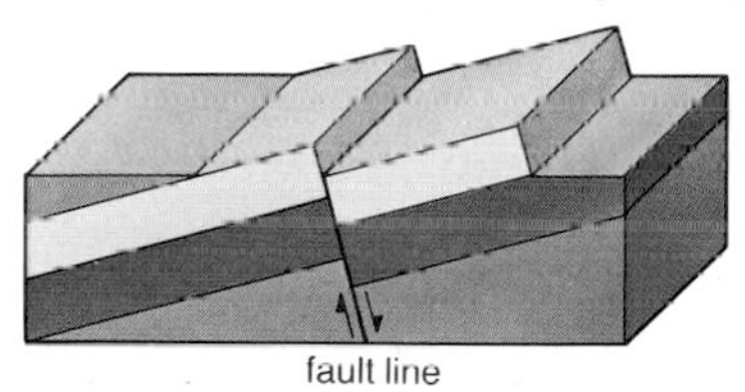

A fault that dips in the direction of the downthrown side is called 'normal'

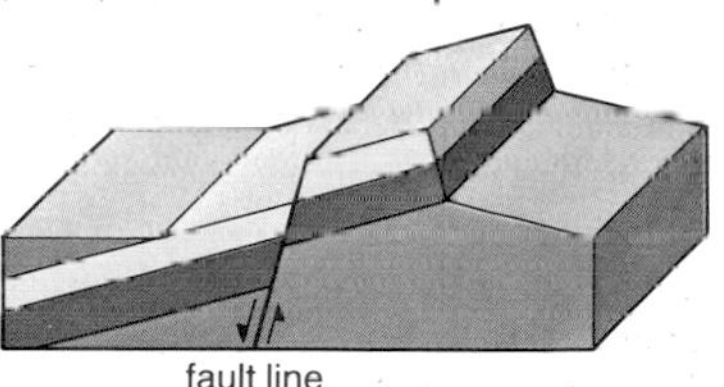

Differing bed dips result in different outcrop patterns.

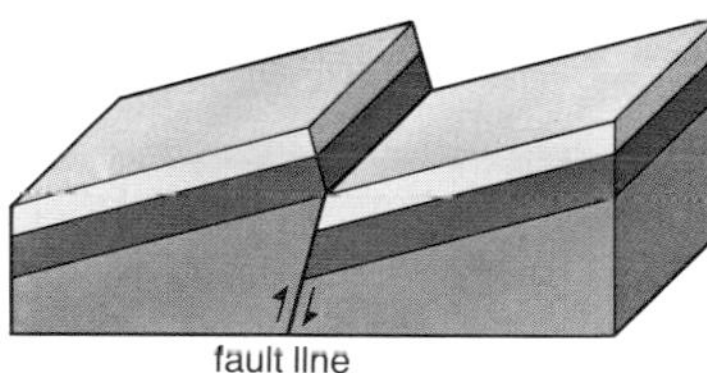

A reverse fault is where the fault plane dips toward the upthrown side.

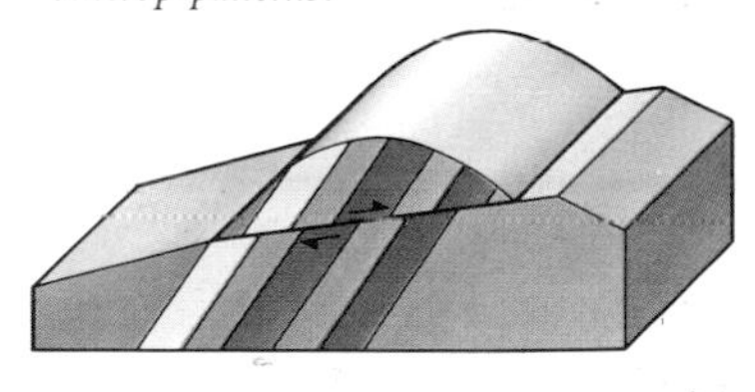

A reverse fault dipping less then 10° is termed a thrust fault.

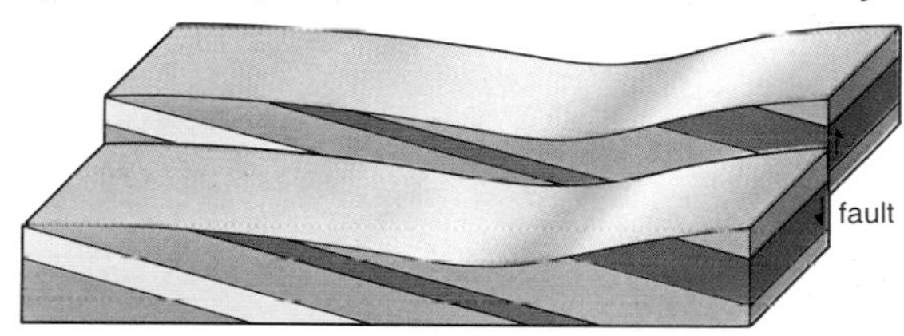

Left: A fault where the fault plane is parallel to the dip of the sedimentary rocks.

Left: A reverse fault in Tunisia. Note the undisturbed strata above the fault showing that it formed prior to these overlying beds.

Metamorphism and metamorphic rocks

The word 'metamorphism', is derived from the Greek **meta** meaning after (denoting a change) and **morphe** meaning shape. The term metamorphism is used to signify the change of one rock type into another new type by the recrystallization of the constituent minerals.

Causes of metamorphism

The changes which occur in metamorphism are in response to conditions not normally encountered at the Earth's surface, such as increases in pressure and temperature. All rocks are stable on the surface of the Earth but, once the physical and chemical conditions change, then so may the rocks, albeit over a long time period. New minerals grow from old, until stability is achieved, and the rock has been transformed or metamorphosed. It is important to remember that these transformations always occur in the solid state; at no stage is the rock ever molten. In some cases, the changes are so subtle as to be virtually undetectable whereas, in other cases, the rock may be totally recrystallized. We can recognize that metamorphic rocks have been derived from igneous, sedimentary, or even other metamorphic rocks, even though their mineralogy and texture have been considerably changed, but their common occurrence does show that these rocks have adapted to new chemical and physical conditions in the solid state to form new and stable mineral assemblages.

In addition to changes of temperature and pressure, there are a number of other factors involved in metamorphism:

1. rocks respond more readily to increases in pressure and temperature than to decreases;
2. a long time period is necessary for metamorphic reactions to run their full course;
3. metamorphic reactions are speeded up by the presence of active chemical agents, such as water, which promote new mineral growth;
4. rocks which are under a strong deforming pressure react more readily than those which are not.

Metamorphic change can be divided into three processes, which usually occur simultaneously over a long period. Completely new minerals may be formed which were not present in the parent rock, although new crystals may replace the original crystals of the same mineral. No overall chemical changes occur; the original constituents are just rearranged. In addition, new textures are produced as new minerals and crystals are formed, and as the minerals are deformed. Most metamorphic changes involve hardly any chemical changes in the overall rock composition. Water and carbon dioxide are, however, frequently added to or removed from rocks during metamorphism.

Types of metamorphism

CONTACT OR THERMAL METAMORPHISM Immediately adjacent to igneous intrusions, the surrounding country rocks become heated up by the temperature of the igneous intrusion. Thus, new minerals are formed and recrystallization takes place. An envelope of metamorphism is produced around the intrusion which is called a **contact** or **thermal aureole.** The metamorphic effects on the country rocks are strongest immediately next to the intrusion but, as the temperature falls rapidly with increasing distance from the igneous body, so does the metamorphism. In this way, the types of metamorphic rocks change across the aureole and eventually reach unmetamorphosed country rocks. We refer to these changes in metamorphic effects as **grade** which is, therefore, a measure of the relative intensity of metamorphism. Within a thermal aureole the grade increases towards the intrusion, so that the rocks immediately adjacent to the intrusion are known as **high-grade** metamorphic rocks, whereas **low-grade** metamorphic rocks are found on the periphery of the aureole.

Basic magmas are hotter than acidic magmas, so it would be logical to expect basic intrusions to develop wider aureoles than acidic intrusions. This is not the case, however, because acidic magmas contain relatively large amounts of water which are expelled into the country rocks carrying heat and speeding up metamorphic reactions.

REGIONAL METAMORPHISM Usually as a consequence of major structural movements of the Earth's crust, such as folding or mountain building, increases of temperature and pressure can cause metamorphism on a regional scale. Rocks produced by regional metamorphism often compose the centres of the large continental masses, and

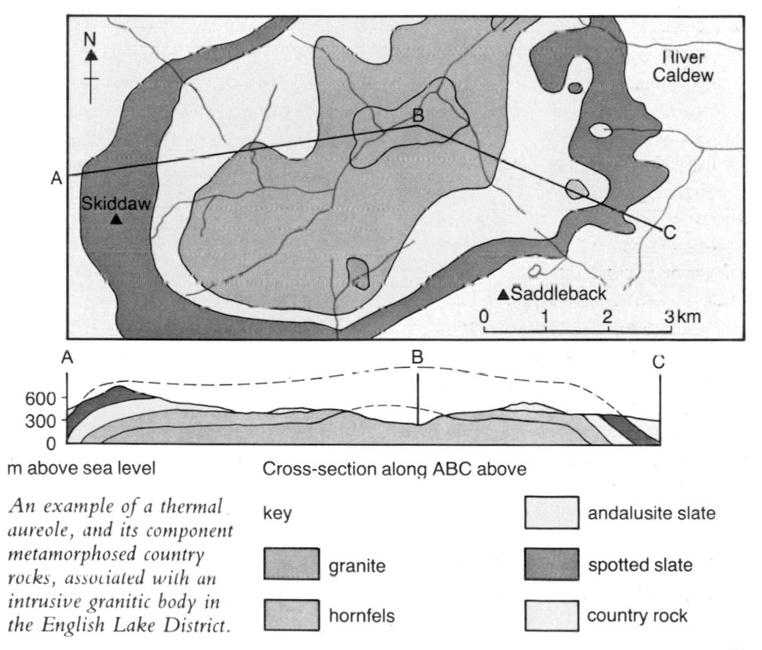

An example of a thermal aureole, and its component metamorphosed country rocks, associated with an intrusive granitic body in the English Lake District.

some of these are the oldest known rocks at around 3500 million years old.

There are other less significant types of metamorphism associated with faulting or deep burial, for example, and it is worth remembering, too, that rocks can be subjected to more than one period of metamorphism during the course of time.

Mineralogy of metamorphic rocks

Many of the minerals found in metamorphic rocks are also common to igneous and sedimentary rocks. Minerals common to all three types of rocks are quartz, feldspar, and muscovite, while biotite, hornblende, pyroxene, olivine, and iron ore are commonly found in igneous and metamorphic rocks. Clay minerals, calcite, and dolomite are found in sedimentary and metamorphic rocks.

There are certain silicate minerals which are either restricted to or found mainly in metamorphic rocks. These are garnet, andalusite, kyanite, sillimanite, staurolite, cordierite, epidote, and chlorite. Of particular interest are the minerals andalusite, kyanite, and sillimanite which all have the same chemical composition Al_2SiO_5. They have different crystal structures and physical properties, however, and are sensitive to pressure and temperature. Usually only one is present in a metamorphic rock because, as a critical temperature or pressure value is reached, one variety transforms into another. For example, with an increase in pressure, andalusite alters to kyanite. These changes can usually be reversed if the conditions revert to their former state.

Metamorphic rocks and their textures

Metamorphic rocks are distinctive in the laboratory and in the field. They are crystalline and are composed of discrete, interlocking crystals of several minerals. Common products of thermal metamorphism are low-grade spotted rocks and high-grade hornfelses (*see* page 163). Regional metamorphism produces low-grade slates (*see* page 165), and medium- to high-grade schists and gneisses (*see* pages 169 and 171). The most striking aspect of any metamorphic rock, however, is its texture. In most cases, it is texture that allows us to identify the different thermal and regional metamorphic rocks.

Regionally metamorphosed rocks are commonly coarser grained than those produced by thermal metamorphism. As a general point, crystals become larger with increasing metamorphic grade. Fine-grained rocks have crystals less than 0.1 mm (0.004 in) across; medium-grained rocks possess crystals between 0.1 mm (0.004 in) and 5 mm (0.25 in) and coarse-grained rocks have crystals greater than 5 mm (0.25 in) in diameter. Thermally metamorphosed rocks, such as hornfelses, are composed of interlocking crystals arranged in a random manner, with no banding or layering. These dense, massive, fine-grained rocks may possibly be confused with other massive sedimentary or igneous rocks, but the field relationships should clarify the origin of the rock. The interlocking crystals of these rocks have roughly the same dimensions in all directions, with adjacent crystal faces often meeting at 120 degrees, characteristic of hornfelses.

Most rocks produced by regional metamorphism nearly always show some degree of **preferred mineral orientation**. This means that the crystals tend to grow in particular directions within the rock. As regional metamorphism involves pressure and temperature increases, new minerals grow at right-angles to the direction of maximum pressure because less 'energy' is expended in promoting growth in this direction. Not all minerals grow or recrystallize in this way, however, but the platy micas (biotite, muscovite, and chlorite) and the long prismatic amphiboles (hornblende, actinolite) always do. Under high pressures, quartz and feldspar may develop a limited degree of preferred orientation. In these cases the rock is said to have a **directional fabric** or simply a **fabric**. These fabrics usually show as visible banding or layering. Platy minerals, such as mica, will adopt a planar preferred orientation giving the rock a **foliation**, while elongated or rod-like minerals, such as hornblende, form a linear preferred orientation so that the rock is said to be **lineated**. If the rock contains both types of minerals, it may exhibit both kinds of fabric.

Planar fabrics or foliations are further described depending on the grain size of the rock in which they are developed. Thus, a **cleavage** or **slaty cleavage** is found in fine-grained rocks such as slate itself (*see* page 166), **schistosity** in medium-grained rocks such as schists (*see* page 169) and **banding** or more accurately, **gneissose banding (gneissosity)** in coarse-grained rocks such as **gneiss** (*see* page 171).

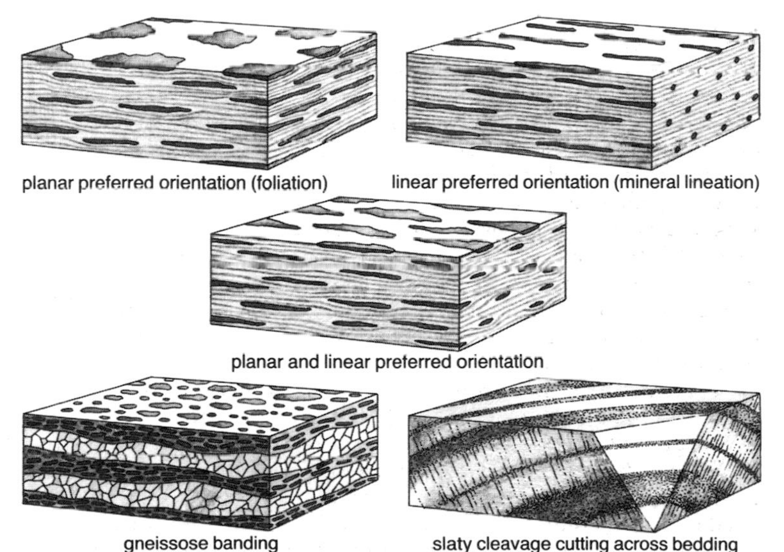

Examples of directional fabrics developed within regional metamorphic rocks. Such fabrics, involving preferred mineral orientation, are unique to these rocks.

Lineations are commonly developed in the rocks called amphibolites (*see* page 157). Regionally metamorphosed rocks possessing a foliation tend to break or split easily in the direction of the foliation. Slate is metamorphosed shale and it cleaves so well into parallel sheets because of the way in which the very fine crystals of mica are arranged, usually cutting across the original bedding. Sometimes, a second cleavage may be developed and superimposed upon the first, which causes the first foliation to become folded or **crenulated**. It is quite common for regionally metamorphosed rocks to be folded, often very complexly, on various scales. In low-grade metamorphic rocks, vestiges of original sedimentary bedding may be recognized, but these are destroyed as metamorphism progresses.

Banding in high-grade rocks should not be confused with sedimentary bedding. Not all regionally metamorphosed rocks possess layering or banding. Some appear as even, medium- or coarse-grained rocks; and rocks such as marbles (*see* page 177) and quartzites (*see* page 173), do not develop visible fabrics.

TEXTURE Most of the terms used to describe the different textures of metamorphic rocks end in '-blast' or '-blastic'. Certain metamorphic rocks, for example, often develop large crystals, such as garnets, in a finer-grained matrix. The large crystals are called **porphyroblasts** and the texture is known as **porphyroblastic**. In porphyroblastic garnet-mica schists, the planar fabric, defined by platy micas, may be clearly deflected and curved around the resistant garnets which grew before the foliation was formed. On the other hand, some porphyroblasts can be shown to have grown after the foliation developed. In this case, the porphyroblasts often enclose crystals of the matrix. These crystals are now referred to as **poikiloblasts** and the texture is **poikiloblastic**.

In certain gneisses derived from coarse-grained acidic igneous rocks such as granites, very large porphyroblasts of feldspar (usually the alkali feldspar, orthoclase) develop. Such porphyroblasts are called **augen** (*auge* is German for eye), and the coarsely banded rocks possessing these augen are called augen gneisses.

Classification of metamorphic rocks

Metamorphic rocks are classified using different criteria from those employed for igneous rocks. These criteria are the environment of metamorphism and the dominant texture of the rocks. For example, we might refer to regional metamorphic rocks based on the environment of metamorphism, and schists based on the dominant texture. The mineralogy of most metamorphic rocks can be rather variable, and greater accuracy can be achieved by prefixing the name of the rock with its important mineral constituents; for example, biotite hornfels, garnet-mica schist, garnet gneiss, or garnet amphibolite.

In some cases, such as in quartzite, the same name is employed for both contact and regional metamorphic rocks. Both types of quartzite are very similar in appearance, and to distinguish them, we need to find out their different metamorphic environments.

Parent rocks	Metamorphic rocks	Description
argillaceous sedimentary rocks	slate (low-grade regional metamorphism)	very fine grained with well-developed cleavage; colour often dark
	phyllite (slightly higher grade regional metamorphism)	fine grained, with crenulated cleavage planes; usually greenish grey
	mica schist (medium-high grade regional metamorphism)	medium to coarse grained; rough, often puckered surfaces parallel to a foliation
	gneiss (high-grade regional metamorphism)	medium to coarse grained; well-developed light and dark bands
mixed sediments or acidic igneous rocks	gneiss (high-grade regional metamorphism)	medium to coarse grained; foliated with quartzofeldspathic and micaceous bands
arenaceous sediments	quartzite (thermal and regional metamorphism)	interlocking quartz grains having granoblastic texture; occasionally micaceous; white
calcareous sediments	marble (thermal and regional metamorphism)	composed of interlocking calcite or dolomite crystals with some calc-silicate minerals; usually pale in colour
basic igneous rocks	actinolite and chlorite schists (low-medium grade regional metamorphism)	green, well foliated rocks with undulating cleavage
	amphibolite (medium to high-grade regional metamorphism)	medium to coarse grained, dark rock; often foliated and banded; mainly composed of hornblende and plagioclase
	eclogite (high-grade regional metamorphism)	medium to coarse grained, dominantly composed of red garnet and green pyroxene
various	hornfels (thermal metamorphism)	many are fine grained, dark, and have no foliation

A simplified table showing the most important metamorphic rocks, with brief descriptions of the original rock types from which they have been derived.

Identifying rocks and minerals

Using the key charts and the colour plates in this section of the book, it should be possible for you to identify most of the major rock types either in the field or from a hand specimen. The rocks have been grouped into the igneous, sedimentary, and metamorphic types which we have described earlier. As well as the illustration of the rock itself, there is an accompanying detailed description and, if its component minerals are relevant to its identification, then these are illustrated, too.

The specimens illustrated have been specially selected to show the appearance of the rocks as they are usually found in their natural surroundings. Please bear in mind, however, that the illustrations do show the rocks and minerals as they would appear in a fresh, unaltered state whereas, in a natural outcrop, the rock may well have suffered millions of years of weathering and alteration. Depending on the type of rock, the weathering may penetrate to a depth of several centimetres and the rock's appearance may be quite different from the fresh face that is shown in the book. To be sure of identifying the rock correctly, therefore, you must, if possible, obtain a sample with a freshly exposed face although you should avoid any unnecessary damage to the outcrop.

The minerals of which a rock is composed give each rock type its distinctive colour and, in many cases, are responsible for its texture. They are particularly useful in the identification of coarse- and medium-grained rocks where it is possible to recognize individual crystals with the naked eye or through a hand-lens. In some cases, we have decided to include illustrations of minerals for fine-grained rocks, too, even though they could only clearly be seen with the aid of a special microscope. This approach does serve, however, to give a better idea of the relationships between these rocks and their coarser-grained equivalents and to explain the overall colour of the rocks.

Identifying the main rock types

The key opposite should enable you to decide if a specimen of rock is igneous, sedimentary, or metamorphic using features which can be seen with the naked eye or with the aid of a hand-lens. Once you have worked out what sort of rock you are dealing with, you should refer to the appropriate key, as indicated, to identify the specimen more accurately. Eventually, you should be able to work out the rock's constituent minerals as well as learning more of its history and formation. The illustrations and descriptions should allow you to confirm your identification.

With a little experience, you should be able to identify most hand specimens relatively easily although there are some more difficult ones which might possess characteristics of more than one group; for example, a quartzite can be either sedimentary or metamorphic. In these cases, their associations in the field should be considered when making an identification.

KEY TO MAIN ROCK TYPES

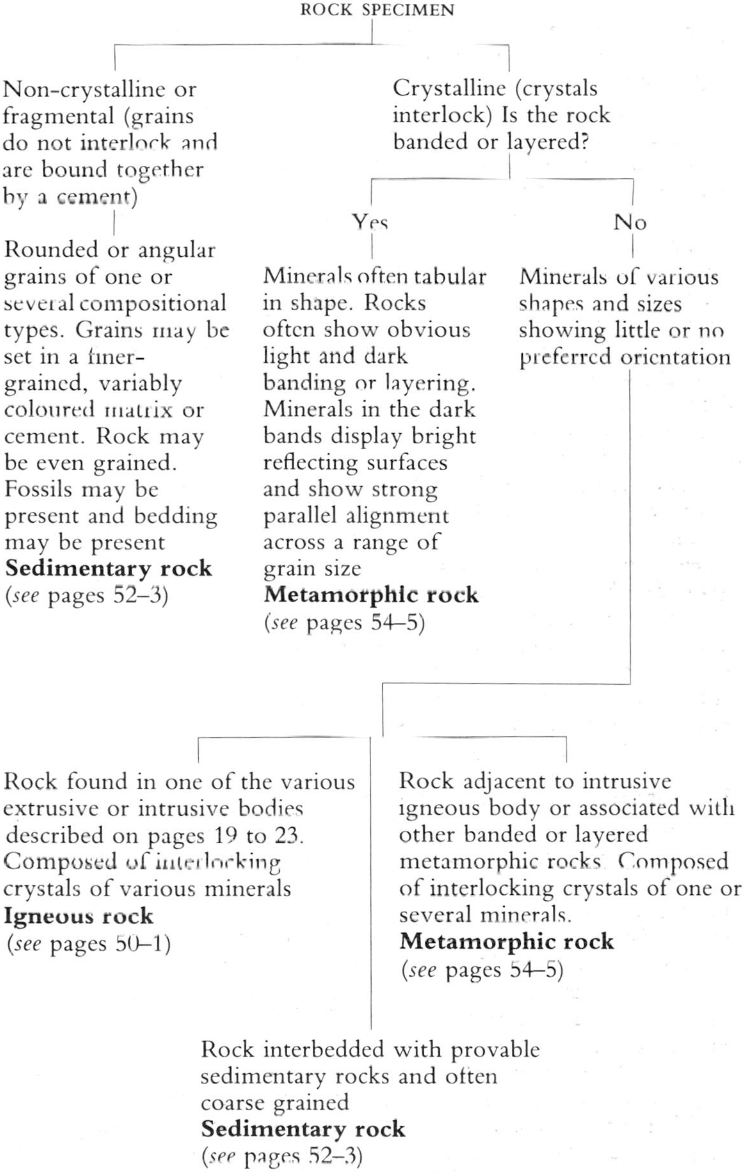

ROCK SPECIMEN

Non-crystalline or fragmental (grains do not interlock and are bound together by a cement)

Rounded or angular grains of one or several compositional types. Grains may be set in a finer-grained, variably coloured matrix or cement. Rock may be even grained. Fossils may be present and bedding may be present
Sedimentary rock
(*see* pages 52–3)

Crystalline (crystals interlock) Is the rock banded or layered?

Yes

Minerals often tabular in shape. Rocks often show obvious light and dark banding or layering. Minerals in the dark bands display bright reflecting surfaces and show strong parallel alignment across a range of grain size
Metamorphic rock
(*see* pages 54–5)

No

Minerals of various shapes and sizes showing little or no preferred orientation

Rock found in one of the various extrusive or intrusive bodies described on pages 19 to 23. Composed of interlocking crystals of various minerals
Igneous rock
(*see* pages 50–1)

Rock adjacent to intrusive igneous body or associated with other banded or layered metamorphic rocks. Composed of interlocking crystals of one or several minerals.
Metamorphic rock
(*see* pages 54–5)

Rock interbedded with provable sedimentary rocks and often coarse grained
Sedimentary rock
(*see* pages 52–3)

KEY TO IGNEOUS ROCKS

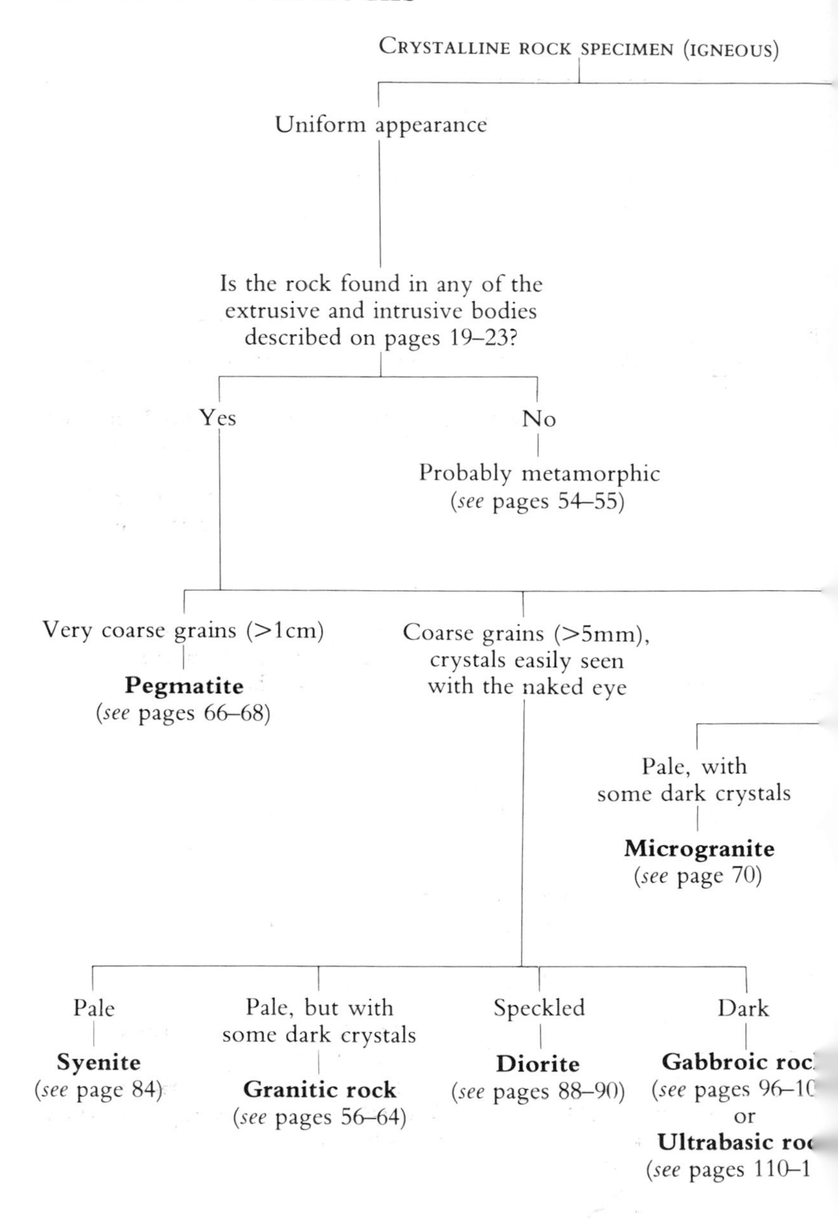

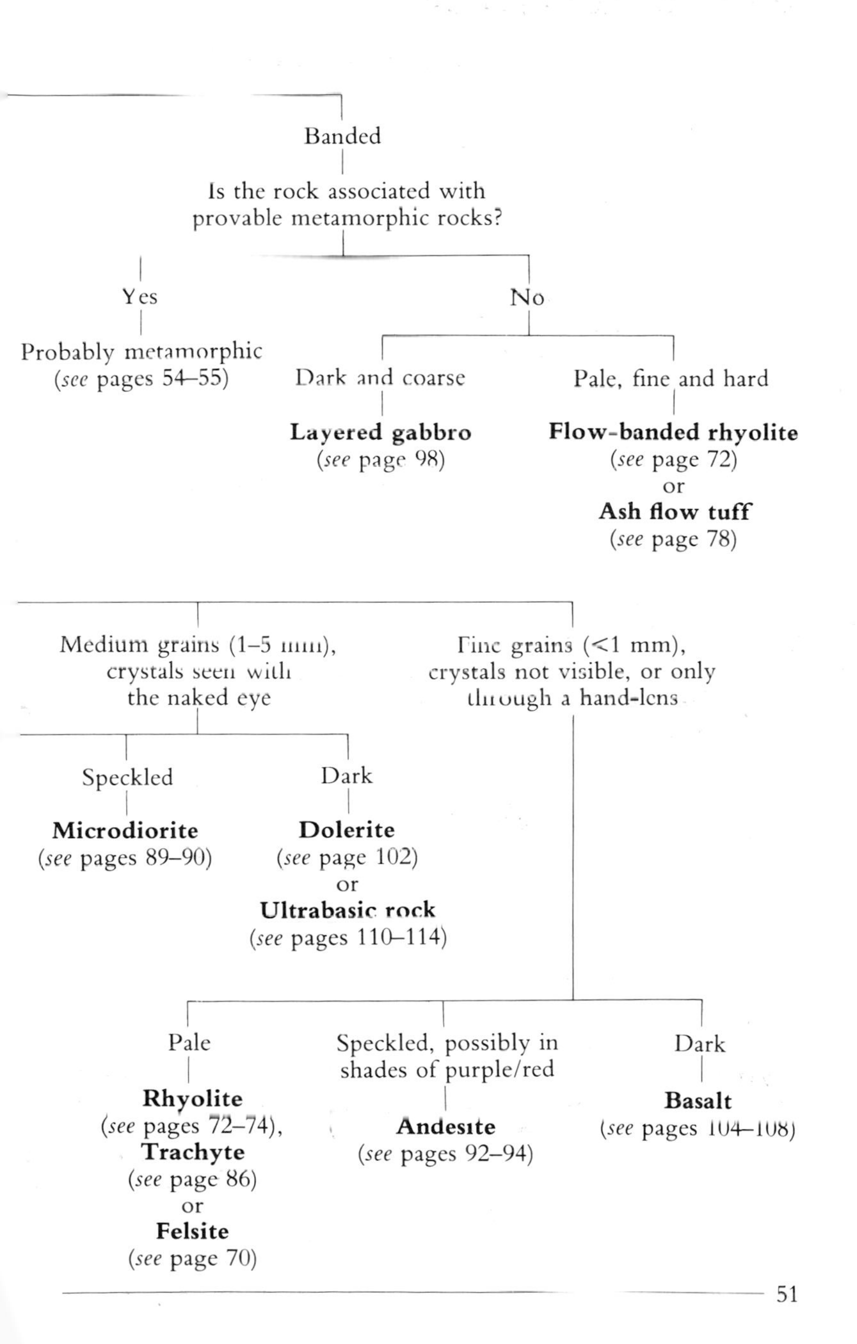
Banded
Is the rock associated with provable metamorphic rocks?
Yes
Probably metamorphic (see pages 54–55)
No
Dark and coarse
Layered gabbro (see page 98)
Pale, fine and hard
Flow-banded rhyolite (see page 72) or **Ash flow tuff** (see page 78)
Medium grains (1–5 mm), crystals seen with the naked eye
Speckled
Microdiorite (see pages 89–90)
Dark
Dolerite (see page 102) or **Ultrabasic rock** (see pages 110–114)
Fine grains (<1 mm), crystals not visible, or only through a hand-lens
Pale
Rhyolite (see pages 72–74), **Trachyte** (see page 86) or **Felsite** (see page 70)
Speckled, possibly in shades of purple/red
Andesite (see pages 92–94)
Dark
Basalt (see pages 104–108)

KEY TO SEDIMENTARY ROCKS

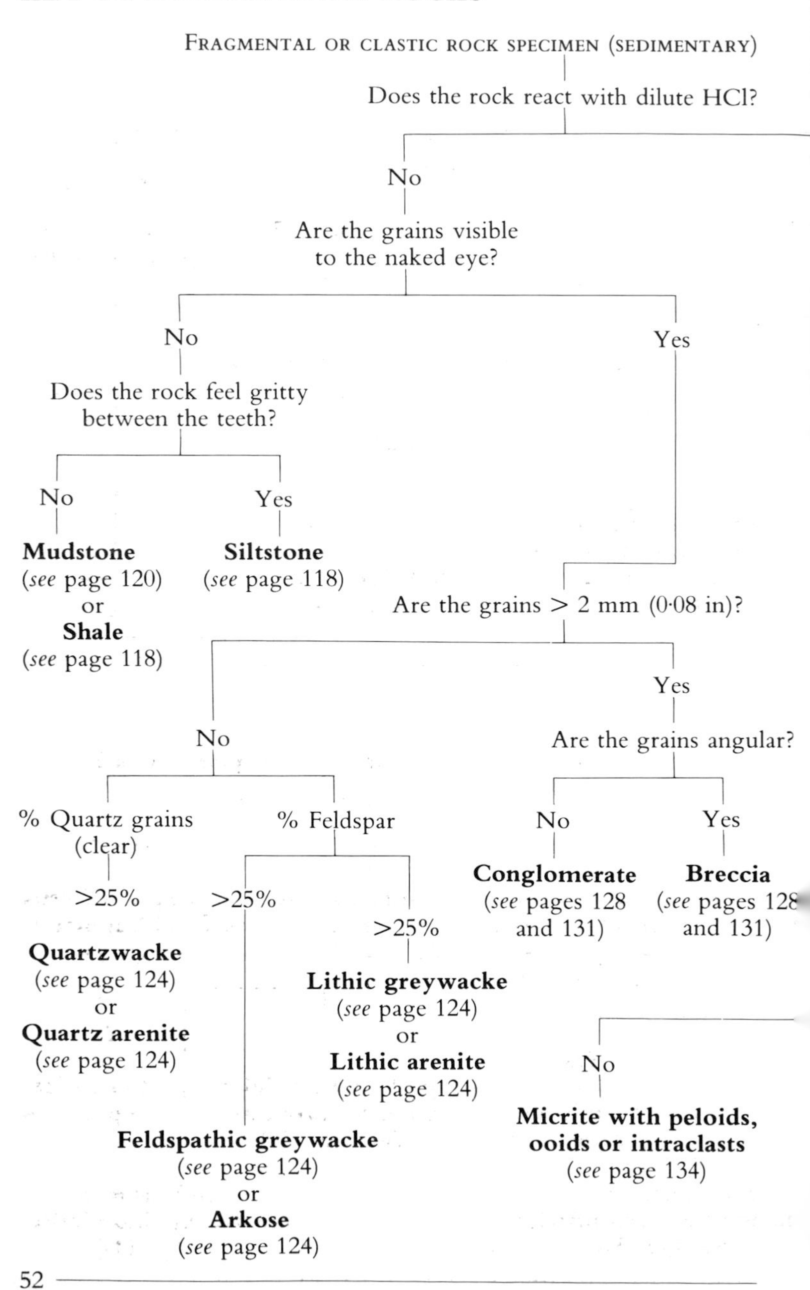

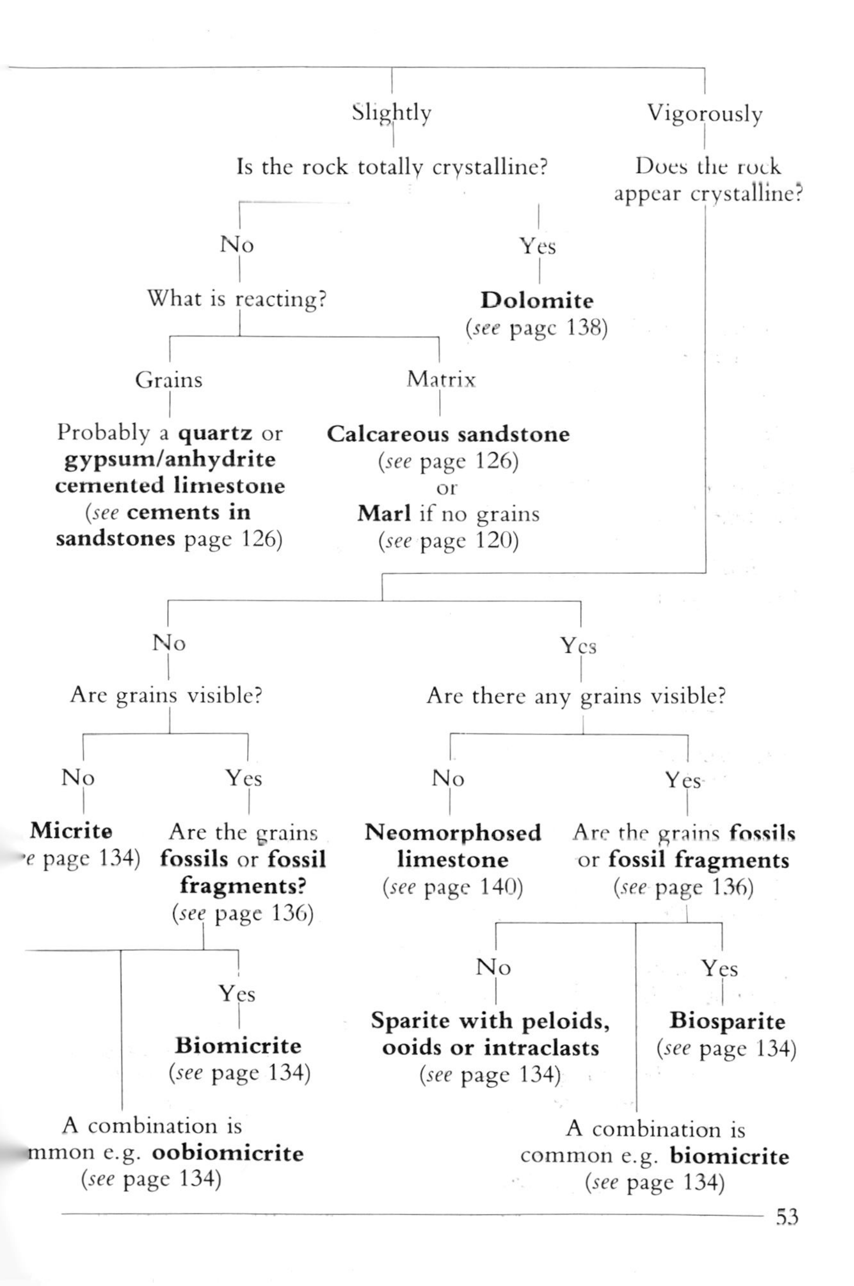
Slightly
Vigorously
Is the rock totally crystalline?
Does the rock appear crystalline?
No
Yes
What is reacting?
Dolomite (see page 138)
Grains
Matrix
Probably a quartz or gypsum/anhydrite cemented limestone (see cements in sandstones page 126)
Calcareous sandstone (see page 126) or Marl if no grains (see page 120)
No
Yes
Are grains visible?
Are there any grains visible?
No
Yes
No
Yes
Micrite
e page 134)
Are the grains fossils or fossil fragments? (see page 136)
Neomorphosed limestone (see page 140)
Are the grains fossils or fossil fragments (see page 136)
Yes
No
Yes
Biomicrite (see page 134)
Sparite with peloids, ooids or intraclasts (see page 134)
Biosparite (see page 134)
A combination is
mmon e.g. oobiomicrite (see page 134)
A combination is common e.g. biomicrite (see page 134)

KEY TO METAMORPHIC ROCKS

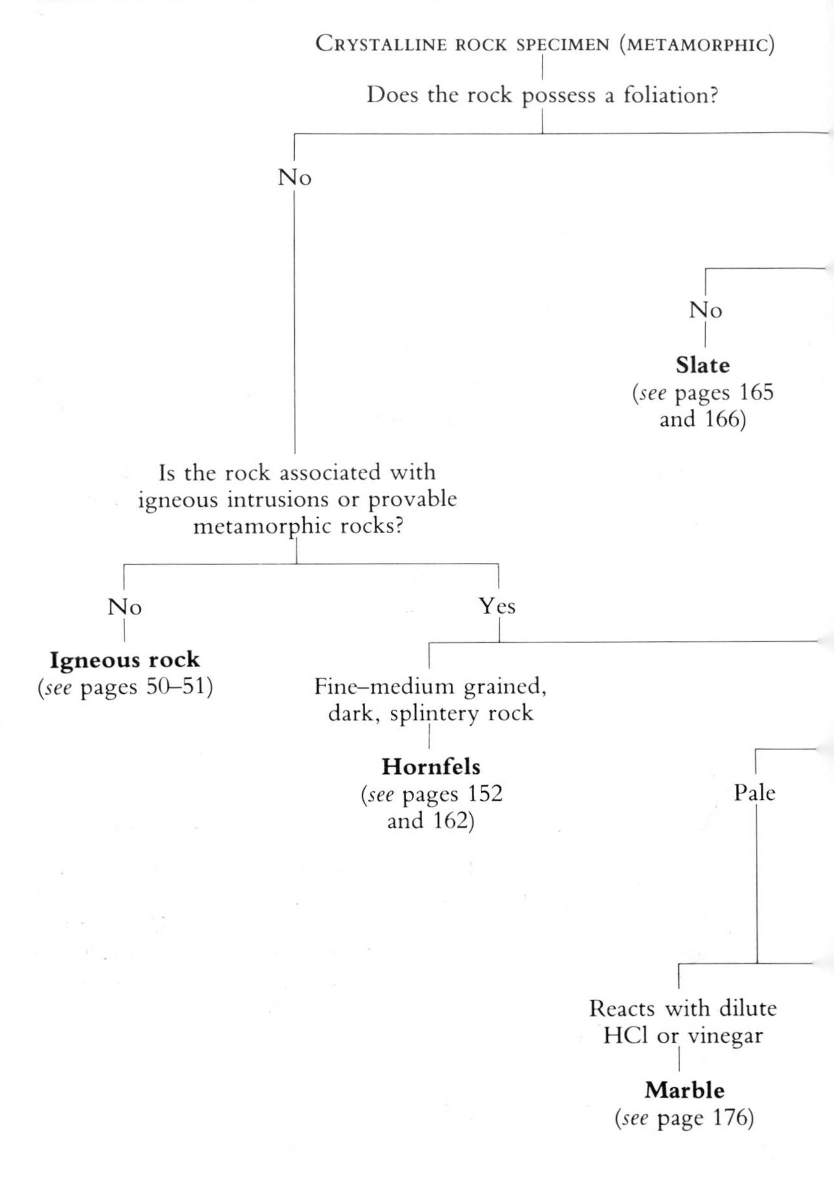

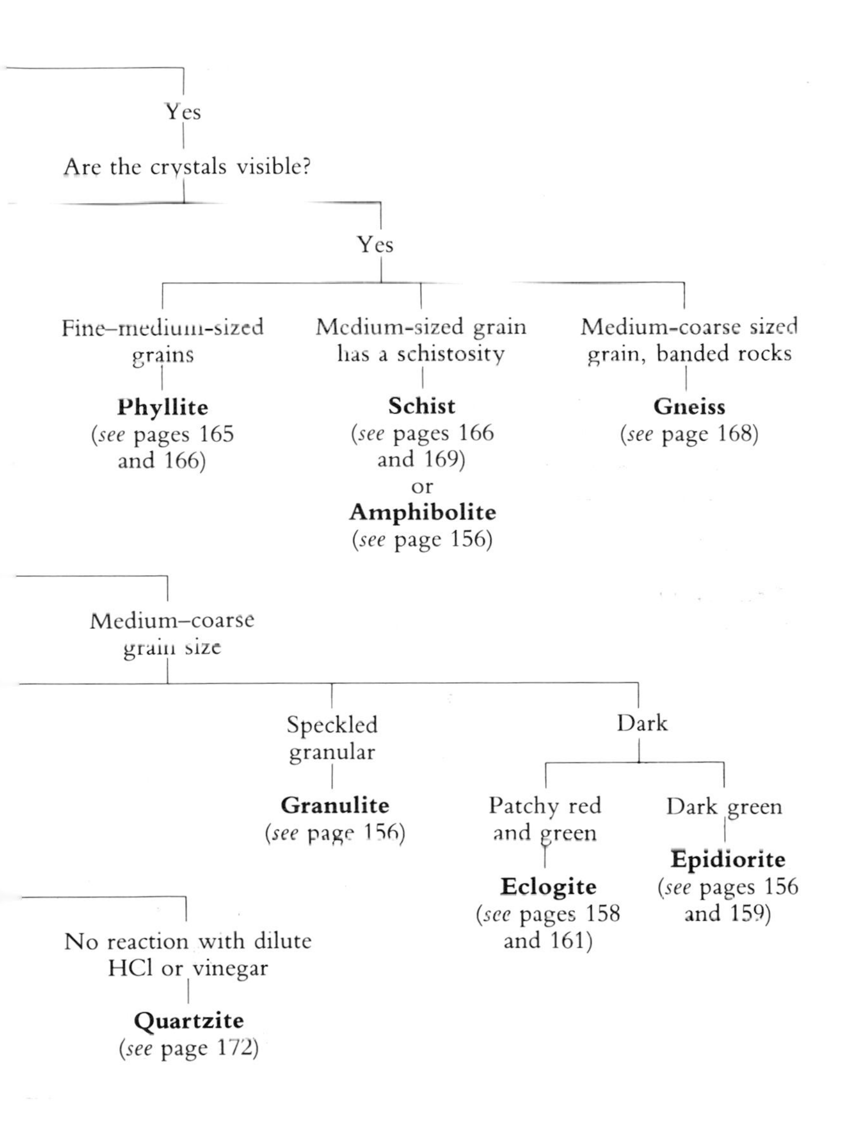
Yes
Are the crystals visible?
Yes
Fine–medium-sized grains
Phyllite
(see pages 165 and 166)
Medium-sized grain has a schistosity
Schist
(see pages 166 and 169)
or
Amphibolite
(see page 156)
Medium-coarse sized grain, banded rocks
Gneiss
(see page 168)
Medium–coarse grain size
Speckled granular
Granulite
(see page 156)
Dark
Patchy red and green
Eclogite
(see pages 158 and 161)
Dark green
Epidiorite
(see pages 156 and 159)
No reaction with dilute HCl or vinegar
Quartzite
(see page 172)

Granites

Granites are intrusive, acid igneous rocks with at least 10 per cent visible quartz. Like the vast majority of igneous rocks, granites are entirely composed of interlocking crystals, and are said to be crystalline. They are generally pale in colour, or leucocratic, ranging from pale grey to pink and red. Dark minerals are never abundant but their presence gives granites a 'speckled' appearance. Sometimes, granites can be porphyritic, with large feldspar crystals surrounded by smaller crystals. Granites are usually coarse grained, often with roughly similar sized individual crystals when they are described as granular. The crystals are randomly arranged, with no regular bands or layers, although feldspar crystals can occasionally show a rough regular alignment owing to the slow flow of the granite magma before it finally solidified.

During the intrusion of the magma, pieces of the country rock often break off and become caught up and preserved as the magma solidifies. Such fragments are called xenoliths and occasionally can be recognized within a granite. These xenoliths commonly have diffuse blurred margins because the surrounding magma was hot enough to cause the outer parts of the rock fragments to melt; in some cases xenoliths may be almost completely melted.

Granitic rocks are usually hard, partly because of the presence of quartz and, when fresh, they are difficult to break. Where the rock has been quarried, the rock faces have abundant protruding angles, and samples can be readily obtained. Be careful to protect your eyes from flying rock chips which can cause painful and even serious injuries. Natural outcrops of granitic rocks tend to have rather rounded outlines and a greater degree of weathering than granite found in quarries. Granitic rocks are rather prone to alteration and can, at times, be relatively soft and crumbly where the rock surface has suffered prolonged exposure to the elements. This is due to the abundance of feldspars which weather and breakdown relatively easily. Here the rock will be simpler to extract but the rock colour and appearance will be modified so you may have to search a little harder for fresh material.

Types of granitic rocks

Granites are usually divided into three broad 'families' or groups. Like the classification of the igneous rocks themselves, these divisions are based on the relative amounts of certain minerals present in the individual rocks. When such rocks are found in the field, however, it is often impossible to distinguish accurately between these minerals. It is probably better, therefore, to describe the rocks as granitic rather than as granites.

The most common silicate minerals in granitic rocks are the feldspars which are used to split up the granitic rocks into three families. The divisions are based on how much alkali feldspar and plagioclase feldspar is present. **Alkali granites** contain only alkali feldspar (such as

muscovite granite
granite
5cm
alkali granite
granite

orthoclase) or, if plagioclase feldspar is present, 66 per cent or more of the total feldspar content is alkali feldspar. **Adamellites** have alkali and plagioclase feldspars in approximately equal amounts. **Granodiorites** may contain only plagioclase feldspar or, if alkali feldspar is present, 66 per cent or more of the total feldspar content is plagioclase.

Occurrence of granitic rocks

Granites are intrusive rocks so that it may take millions of years of erosion before the overlying country rocks are worn away and the granite is exposed at the surface. Granitic rocks are commonly found as huge batholiths, and the exposure of such a large area of hard rock has a profound effect on the landscape. When exposed, batholiths generally form large upland areas, usually with a thin covering of acidic soil supporting sparse vegetation. For example, moorland areas, such as Dartmoor, in south-west England are excellent examples of such areas. Here the batholith is not completely exposed; rather, there are extensive tracts of moorland with occasional upstanding blocks of rock capping hills. In south-west England, these blocks are called **tors**, and are commonly associated with granitic rocks. A tor looks like a pile of rock slabs. During cooling and solidification of a granitic magma, the rock contracts, and regular sets of fractures or joints develop. Weathering proceeds readily along the joints, ultimately reducing the rock mass to a pile of slabs.

Granitic rocks are common worldwide, and are noted from such localities as the Lake District in northern England; New Hampshire, Colorado, and Virginia, USA; western Australia; the Sudan in Africa, and many other places.

Formation of a granite tor

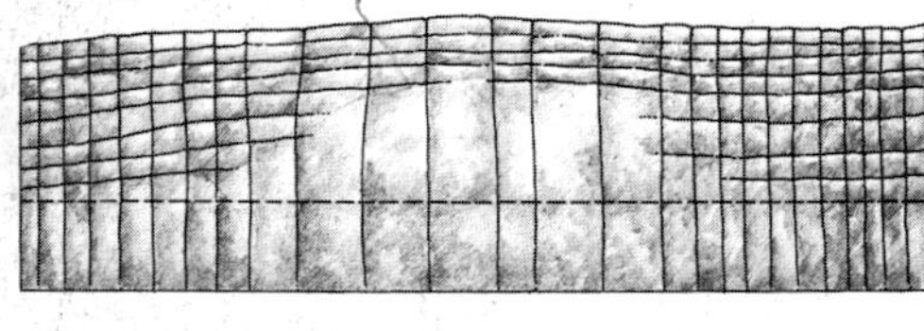

1 Freshly cooled granite showing fractures and joints which formed during the cooling process.

2 Weathering proceeds to decompose the rock along these joints and fractures.

3 Eventually all the decomposed rock is weathered away leaving the distinctive upstanding tor.

granodiorite

granodiorite
with xenoliths

xenolithic granite

granite with xenoliths

5cm

granite

Essential minerals

QUARTZ All granitic rocks contain at least 10 per cent of visible quartz. Although it is colourless and transparent, quartz often appears as grey rounded crystals, randomly distributed between the larger feldspar crystals. Characteristically, it has a glassy appearance or vitreous lustre and does not possess a cleavage. Quartz is extremely resistant to weathering, so it will usually appear fresh and unaltered. Quartz has a hardness of 7 on the Mohs' scale. Occasionally, it may be faintly tinged with colour. Sometimes it is found infilling veins or cavities within the granitic rock.

FELDSPAR Feldspars comprise the greater part of a granite rock and are usually more abundant than quartz. They are generally either white or creamy pink and give the rock its overall colour. They are easily distinguished from quartz by being slightly softer (they can be just scratched with a penknife) and by having cleavage planes. Unfortunately, the different types of feldspar can be very difficult to distinguish in hand specimens. The alkali feldspar, **orthoclase**, can show pink-cream shades, while plagioclase feldspar is usually white. Do not rely too heavily on colour, though, because this is due either to the presence of impurities or to alteration. The abundance of orthoclase and plagioclase feldspars will vary depending on the type of granitic rock. Another alkali feldspar, **microcline**, may also be present in granitic rocks. Again, it has very similar properties to the other feldspars and is indistinguishable in hand specimen, apart from the green variety called **amazonstone.** A useful guide is that the most common alkali feldspar in granitic rocks is generally orthoclase, while microcline is often the most common alkali feldspar in pegmatites and veins associated with these rocks. Alkali and plagioclase feldspars are often found together as **perthite** or perthitic feldspar, which often has a diffusely banded appearance. On a weathered surface of a granitic rock feldspars commonly develop a paler colour and become crumbly.

BIOTITE MICA This is one mineral which gives granitic rocks their dark 'speckles'. It occurs as randomly distributed, small, dark plates or flakes, and possesses one perfect cleavage. The cleavage planes have a well-developed vitreous lustre. Biotite is quite soft (hardness $2\frac{1}{2}$) and scratches easily with a penknife and with a fingernail.

MUSCOVITE MICA Muscovite or white mica is similar in distribution and properties to biotite but it is colourless and may be easily overlooked.

HORNBLENDE Hornblende is not usually abundant in granitic rocks, accounting for no more than 10 per cent of the mineral content. It is dark green to black and often forms small, elongated (prismatic) crystals. Together with biotite, hornblende gives granitic rocks their dark patches. With a hand lens, it can be seen that hornblende has two inclined sets of cleavage planes and the lustre, although vitreous, is not as strong as that of biotite. It is also much harder (6) and cannot be scratched so easily.

quartz

orthoclase

plagioclase

biotite

hornblende

microcline

muscovite
5cm

granite

Accessory minerals

Many accessory minerals are found in granitic rocks and only the most common will be described. In general, the accessory minerals are less common than the essential minerals and account for only 2 to 3 per cent of the rock. They can be rather small and easily overlooked.

TOURMALINE It is usually black although tourmaline can also be brown and dark blue. It often occurs as fairly large, prismatic crystals, up to 1 to 2 cm (½ in) across, with rounded triangular cross-sections. Crystal faces, if seen, often possess parallel lines or striations. It does not have a well-developed cleavage and is hard (7½) so it cannot be scratched by a penknife.

APATITE This occurs as tiny needles or prismatic crystals and can easily be missed. Apatite is often colourless, but can develop pale greens and blues. It does not show a cleavage and can be scratched with a penknife, having a hardness of 5. Well-formed crystals may show six-sided cross-sections.

ZIRCON Zircon is found as small, colourless to pale-brown prismatic crystals with square cross-sections. They do not usually show a cleavage and, having a hardness of 7½, are not easily scratched.

SPHENE Sphene occurs as small, brown to greeny yellow wedge- or diamond-shaped crystals with a fairly well-developed cleavage. They can be scratched with a penknife (hardness 5½).

PYRITE Sometimes this is cube-like, or may develop as a thin layer on the surfaces of joints. It has a pale, brassy yellow colour. When subjected to weathering, pyrite may tarnish, the brassy yellow colour becoming deeper and other iridescent colours developing. It has a brown-black streak, a metallic lustre, and a hardness around 6½, so it can just be scratched with a penknife.

CHALCOPYRITE Chalcopyrite is superficially similar to pyrite although well-formed crystals develop four triangular faces (tetrahedron). The brassy yellow colour is slightly darker and may develop a purplish tarnish. It is readily distinguished from pyrite because of its lower hardness (3½), making it easy to scratch with a blade. It possesses a similar metallic lustre.

MAGNETITE This is often found as small, black, eight-sided (octahedral) crystals giving a black streak or as small, granular aggregates. It does not show a cleavage and often displays a strong metallic lustre. It has a hardness of 6, so it may just be scratched by a penknife.

granite

5cm

apatite

zircon

sphene

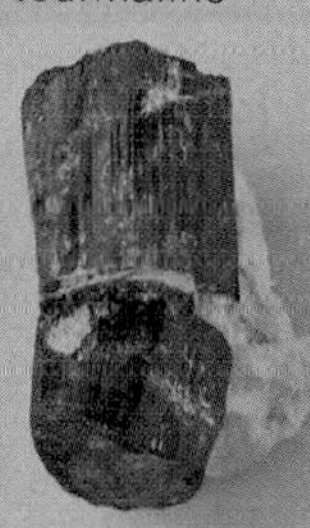
tourmaline

pyrite

granular aggregate
of magnetite

chalcopyrite

Alkali granites

Alkali granites are usually leucocratic with up to 30 per cent visible quartz, and a variable amount of scattered dark grains of biotite mica, along with some colourless muscovite flakes. The majority of the rock is composed of feldspars, of which alkali feldspar (orthoclase) predominates. Occasionally, hornblende may be present. Good examples include the Conway alkali granite of New Hampshire, USA, alkali granites from the Channel Islands, certain examples from the Western Red Hills on the Isle of Skye off the west coast of Scotland, and examples from western Australia and southern Brazil, especially around Rio de Janeiro.

Adamellites

These granitic rocks are still classed as leucocratic, but are slightly darker than alkali granites because of the slight increase in the dark minerals, biotite and hornblende. Visible quartz is up to 30 per cent. Feldspars are still dominant but alkali feldspar and plagioclase feldspar are found in almost equal quantities. It is the colour of the orthoclase and the amount of dark minerals which govern the overall colour and appearance of these granitic rocks. A well-known example is found at Shap Fell in the English Lake District where the adamellite is characterized by large crystals of orthoclase feldspar (30 per cent), with smaller white plagioclase feldspar crystals (30 per cent) and grey glassy quartz crystals (25 per cent). Microcline can be present but is difficult to identify accurately in hand specimen. The chief dark mineral is biotite with lesser amounts of dark-green hornblende, with zircon, apatite, magnetite and pyrite as common accessory minerals. Other examples are from Devon and Cornwall in south-west England, the Sierra Nevada batholith of California, and the type locality of the Adamello complex in northern Italy.

Granodiorites

These are leucocratic although generally they are a little darker than adamellites due to the presence of slightly more biotite and hornblende, of which the hornblende is more abundant. Visible quartz is still up to 30 per cent, and feldspars remain the dominant minerals, although plagioclase feldspar is much more common than orthoclase. Some muscovite flakes may be present. Apatite and sphene are among the common accessories. Not only are granodiorites the most common of all the granitic rocks but they are possibly the most common intrusive igneous rocks of all. Examples occur in parts of the Sierra Nevada batholith in California, the massive granitic batholiths comprising the mountain ranges of western North America, and the Strontian granodiorites in western Scotland.

granite
alkali granite
granodiorite
granodiorite
granite
5cm

Pegmatites

Pegmatites are extremely coarse grained rocks which may yield many excellent mineral specimens. Most crystals exceed 2.5 cm (1 in) across; certain crystals even reach 1 to 2 m (3 to 6 ft) across, and even gigantic crystals of 15 m (50 ft) have been found in pegmatites. Most pegmatites have the typical crystalline appearance of igneous rocks, with randomly distributed interlocking crystals. By pegmatites, geologists often mean a very coarse-grained igneous rock containing the same minerals that make up a granitic rock. Strictly, such a rock should be called a granite pegmatite because there are also pegmatites which have combinations of minerals more typical of other igneous rocks such as diorite or gabbro. Usually, however, pegmatites are intimately associated with intrusive granitic rock bodies, commonly taking the form of veins, small dykes, sills, or irregular masses, which are very often situated at the margins of the larger intrusive bodies. Not every intrusive body has associated pegmatites, however.

Formation

The formation of a pegmatite is rather complex. During the slow cooling of an intrusive igneous magma, hot gases and certain hot fluids containing dissolved elements become concentrated within parts of the igneous body, especially within fractures of the country rock produced during intrusion. Under these conditions of temperature and pressure, large crystals are able to grow. Several rare elements are concentrated in pegmatites, along with certain exotic minerals which are rare elsewhere.

Apart from coarse grain size, another characteristic feature of granite pegmatites is the intergrowth of feldspar and quartz crystals producing **graphic** or **runic texture** (see page 69), in which the feldspar is white and the quartz is grey and glassy. As pegmatites often develop within fractures, the minerals commonly grow inwards into the open space, becoming progressively larger as they do, and giving a characteristic appearance to the rock. One further feature is the occurrence of regular zones of different minerals or mineral combinations within the pegmatite body, although this feature can often be difficult to recognize.

Occurrence

Pegmatites are quite abundant and are often associated with intrusive igneous bodies, such as around the Cornish granites in south-west England. The pegmatites of the Spruce Pine district of North Carolina, the Middletown district of central Connecticut, the Keystone district of south-western South Dakota, and the Winnipeg River area of Manitoba, Canada, are especially worth a mention.

Pegmatites are most commonly found in dyke- and sill-like bodies, and they exert no real influence on the landscape. When you are collecting specimens from a pegmatite, take great care because indiscriminate hammering will invariably damage and destroy good crystals, and ruin

examples of pegmatites

5cm

the outcrop for future visitors. A small cold steel chisel 0.5 to 1 cm ($\frac{1}{4}$ to $\frac{1}{2}$ in) across will prove useful when extracting specimens, because single crystals can be removed without damaging adjacent crystals.

Essential minerals

These are the same as those found in the acid intrusive rocks (*see* page 60).

Accessory minerals

Tourmaline, apatite, pyrite, chalcopyrite, magnetite, sphene, and zircon are frequent accessories. Tourmaline can often acquire striking red and green colours.

LEPIDOLITE MICA Its physical properties are very similar to those of muscovite (*see* page 60), but lepidolite often develops a pale lilac colour. This is due to the presence of lithium, for which this mica is mined.

SPODUMENE This mineral is one of the pyroxenes and is also rich in lithium. It can grow to considerable size – spodumene crystals up to 15 m (50 ft) across have been found in the Etta pegmatite near Keystone, South Dakota. It has two cleavages and is quite hard ($6\frac{1}{2}$), and may just be scratched with a penknife. It shows shades of transparent pink, violet, green, or white, and has a white streak.

BERYL This mineral is rich in the element beryllium for which it is mined. It can form large, prismatic crystals which do not show cleavage. It is commonly pale green in colour, but gem varieties can be transparent dark green (emerald) or pale blue to green (aquamarine). Beryl has a white streak. The diagnostic feature is its hardness, which is 8, thus making it very difficult to scratch. Well-formed crystals may have regular, six-sided (hexagonal) cross-sections.

TOPAZ This mineral is often colourless and transparent, but can be pale blue, yellow, or yellowy brown. It has a white streak, and one well-developed cleavage which is parallel to the base of the crystal. An individual topaz crystal is quite heavy; in addition, it has a hardness of 8, making it extremely difficult to scratch.

FLUORITE Often found as cubic crystals. When pure, fluorite is colourless and transparent, but impurities produce yellow, green, blue, and violet hues. It possesses four cleavages and is fairly soft (4). Fluorite may be confused with calcite, but calcite is even softer ($2\frac{1}{2}$) and does not occur as cubes.

CASSITERITE Cassiterite, or tin oxide, can be yellow, reddy brown, or browny black. Well-formed crystals have a pyramidal shape, but they can also be found in massive aggregates. Cassiterite has a hardness of 6 to 7 and so it is not easily scratched. Individual crystals are very heavy. The lustre of cassiterite crystals is very bright and metallic.

quartz
microcline
perthite
muscovite
biotite
5cm
graphic texture
beryl pegmatite
lepidolite
fluorite
tourmaline
cassiterite
2cm
beryl
spodumene
topaz

Granitic rocks commonly found as dykes and sills

FELSITE This is a general term for fine-grained acid igneous rocks occurring in dykes and sills. Often the component crystals cannot be distinguished, even using a hand lens, and the rock appears to be compact, pale pink to grey, and very fine grained. Felsites comprise of quartz and alkali feldspar which consist of complexly intergrown crystals, and result from **devitrification** (*see* pages 74 and 76).
MICROGRANITE As the name implies, this rock is just a finer-grained version of granite. Perhaps, therefore, they should be referred to as micro-alkali granites, micro-adamellites, and micro-granodiorites. They are light-coloured rocks, with randomly distributed, small, black crystals (usually biotite) giving the rock a speckled appearance. Although microgranites are finer grained than granites, the individual crystals are still visible to the naked eye. The essential and accessory minerals of microgranites are those that are found in the granitic rocks (*see* page 60).
PORPHYRITIC MICROGRANITE Occasionally, large crystals about 0.5 to 1 cm ($\frac{1}{4}$ to $\frac{1}{2}$ in) are developed within the microgranite. These are surrounded by much smaller crystals giving the rock a porphyritic texture. Frequently, these phenocrysts are of quartz and feldspar, and the rock is called a **quartz porphyry** or **quartz-feldspar porphyry**. More generally, they are termed **granite porphyries**, or simply **porphyrites**. Less often, dark minerals, such as biotite and hornblende, may form phenocrysts. The rocks are leucocratic and, around the large crystals, the groundmass comprise the same minerals but these are usually too fine to be distinguished accurately. The mineralogy is the same as that of a microgranite. There is a common variety of porphyritic microgranite in which the quartz and alkali feldspar are very complexly intergrown on a fine-grained scale which is not visible with the naked eye. This complex intergrowth is termed **granophyric texture**. Such rocks are pale pink to white in colour and called granophyres after their characteristic texture. These are frequently found as sills and dykes, but can form relatively large intrusions.

Occurrence

All these rocks are hard, so the dykes and sills may form upstanding linear features in the landscape as the surrounding softer rocks have been more easily eroded. Examples include the quartz porphyry dykes associated with the granitic batholiths in south-west England and certain granites in the Criffel-Dalbeattie region of western Scotland, although dykes and sills of these rock types are common worldwide. It is generally assumed that the dykes and sills were introduced into fractures produced by the intrusive granitic rocks, with which they are usually associated, and that all these rocks are of approximately the same age.

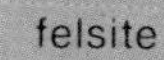

pink felsite

felsite

pink microgranite

microgranite

porphyritic microgranite

quartz feldspar porphyry

feldspar porphyry

5cm

Rhyolite

Rhyolite is the general name given to several different types of extrusive acid igneous rocks, and it may be better to refer to them as rhyolitic rocks. Rhyolites are the extrusive equivalents of the intrusive granites and they each have the same mineral and chemical compositions. Unfortunately, rhyolites are extremely fine grained and, usually, minerals cannot be accurately identified even with a microscope.

Rhyolites are leucocratic, in shades of white, grey, pale green, red, or brown. They are not usually heavily weathered and, characteristically, their thin weathered surface is either white or pale cream, while the fresh interior surface may appear quite dark because of the fine grain size. Rhyolites are very hard rocks and break with a conchoidal fracture like glass. Take great care to protect your eyes when collecting rhyolites because their extreme hardness causes them to splinter into sharp, jagged fragments when hammered.

Structures

Rhyolites are not always completely structureless. Occasionally they show well-defined banding or layering which may be fairly straight over short distances but is often intricately folded and convoluted. This is called **flow banding**. The visible layers are caused by slight changes in grain size and colour, slight differences in mineral composition or to the concentration of gas bubbles in the original lava. Acid lavas are characteristically viscous and flow very slowly. The tortuously folded flow banding displayed by rhyolites reveals all the contortions that this 'stiff' lava went through as it flowed and before it finally solidified.

Rhyolites are very fine-grained rocks, but they sometimes contain larger phenocrysts and they are then called porphyritic rhyolites. Sometimes, these crystals are randomly arranged, but they may be confined to specific bands within the rhyolite and can also be aligned (*see* page 75). In this case, the phenocrysts had crystallized out within the lava before it solidified completely, and the subsequent flowage of the lava caused them to become aligned. The phenocrysts are commonly sanidine feldspar and quartz.

There may be small, round holes or vesicles in rhyolites which represent gas bubbles within the original lava. These vesicles may become filled with minerals, such as quartz and feldspar, and by white minerals called zeolites, and then the vesicles are called amygdales.

Small spherical bodies, 0.5 to 1 cm ($\frac{1}{4}$ to $\frac{1}{2}$ in) or larger, are frequently present in rhyolites, either randomly distributed or restricted to diffuse bands. These are called **spherulites** and the rocks are described as being **spherulitic**. In addition, rhyolites, and indeed granites, occasionally contain larger orb-like bodies, several centimetres across which can be separated or joined. Some have a core, and internal concentric banding can often be recognized. On weathered surfaces, the outer, fine-grained shell of these orbs is pale grey, while the inner part is somewhat darker

orbicular rhyolite
5cm

porphyritic rhyolite

'flow-banded' rhyolite

'flow banded' rhyolite

weathered rhyolite

and rougher because of the slightly coarser grain size. This **orbicular texture** may have resulted from the reaction of the magma or lava with a foreign body or xenolith (*see* page 73).

GLASS All extrusive rocks, irrespective of their composition, have suffered rapid cooling, which causes the rocks to be fine grained. If the lava cools very quickly, it will solidify as a glass because there was not enough time for crystals to form. Natural glass does not contain any crystals but it is not stable and, over a long period (up to several million years), very small, slender, fibrous crystals called **microlites** gradually form and replace the glass. This process, called **devitrification**, starts from centres within the rock and spreads outwards, the small slender crystals, of varieties of quartz and feldspar, radiating out from these centres. On a flat surface of the glass, these radiating aggregates appear circular but they are actually spherulites and may have hollow centres.

Mineralogy

Because of their fine grain size, the mineral composition of rhyolites can only be determined accurately using sophisticated chemical techniques. Only the phenocrysts in porphyritic rhyolites can be identified in the field. The mineralogy of rhyolites is essentially the same as that of the granitic rocks and the microgranites. As with the other acid rocks, quartz and feldspar are the most common. Varieties of quartz, such as cristobalite and tridymite, which only crystallize at relatively high temperatures, are present. High-temperature alkali feldspars are also found, such as sanidine, along with orthoclase and plagioclase feldspar. Some dark minerals occur, usually as phenocrysts, especially biotite and, more rarely, hornblende, along with other minerals including pyroxenes (augite, hypersthene, and olivine). Garnet may also form phenocrysts in rhyolites. Accessory minerals include pyrite, zircon, topaz, fluorite, and apatite.

Occurrence

Rhyolites are generally found as flows, sills, and dykes, and occasionally as plugs where the original, highly viscous lava blocked the throat of an ancient volcano. They never form widespread deposits because the viscous lava can only flow for short distances. Owing to their hardness and resistance to weathering, rhyolites often form jagged, angular, upstanding rock masses and these features, combined with their pale colours, make rhyolites relatively easy rocks to identify in the field.

Rhyolites are relatively common rocks. Sodium-rich rhyolites are found on the islands of Sardinia and Pantelleria in the Mediterranean; Kenya and Nigeria in Africa; the Lahn district in Germany; Mono Lake in California; Sugarloaf Hill in Arizona, and in Iceland. Other examples of rhyolites come from Snowdonia in North Wales; the English Lake District; Devon and Cornwall in south-west England; Yellowstone National Park in north-western Wyoming and Mount Rogers in southern Virginia. Spherulitic rhyolites are noted on the north coast of Jersey in the Channel Islands.

quartz
orthoclase
plagioclase
biotite
'flow-banded'
porphyritic rhyolite
5cm
rhyolite
apatite
pyrite
garnet

Glassy rocks associated with rhyolites

Obsidian, pitchstone, and pumice are acidic rocks often associated with rhyolites. They are all extremely fine grained; no crystals are usually visible even under a hand lens. These rocks all have the same overall composition as rhyolites and granitic rocks, although it is hard to imagine a greater contrast between obsidian, a hard, black, glassy rock which splinters into sharp, jagged fragments, and pumice, a white rock, full of holes, and which is so light it will float on water.

OBSIDIAN Also known as Iceland agate, obsidian is a natural acidic glass. It is black or deep brown when fresh, although it can be pale grey when weathered. Obsidian contains no crystals or bubbles. When hammered, it breaks with a conchoidal fracture. The fracture surface is relatively smooth but it is broken by several curving concentric ridges, centred around the point of impact. Obsidians are derived from rapidly cooled rhyolitic lavas. Over a long period of time obsidian devitrifies forming patches of pale-grey, feathery crystals of quartz and feldspar, and the popular name given to this rock is snowflake obsidian. Obsidian may also show flow banding and spherulites (*see* page 187).

PITCHSTONES These are black, dark-red or dark-green glassy rocks, although they have a duller lustre than obsidian. They generally have suffered more devitrification and are more crystalline than obsidian and are often porphyritic. Phenocrysts of quartz, sanidine (an alkali feldspar), plagioclase feldspar, and light-green pyroxene can be identified and microlites are also present. Pitchstones develop cracks and flow structures.

PUMICE This is a glassy rock containing many gas holes which make it very light. Although, strictly, its composition ranges from basic to acid, pumice is usually regarded as acidic. It is a pale rock and represents the solidified froth or crust which has formed on the surface of viscous, acidic lavas containing much gas. Occasionally, phenocrysts of sanidine, plagioclase feldspar, and quartz may be recognized.

Occurrence

These rocks are usually found as lava flows and chilled marginal zones in dykes and sills. As obsidian and pitchstone are relatively hard, they form upstanding areas, although they never achieve any large lateral extent. Pumice is usually found as loose blocks, associated with lava flows and volcanoes.

Obsidian is found on Lipari Island in the Mediterranean; Glass Mountain in Sisikyou County, California; Obsidian Cliff, Yellowstone National Park, Wyoming; Glass Butte, Oregon; Big Obsidian Flow, Newberry Caldera, Oregon, and Mount Hekla in Iceland. Examples of pitchstone include those on the Hebrides off western Scotland (Sgurr of Eigg) and at Silver Cliff in Colorado. Pumice is generally associated with rhyolite and explosive volcanic activity, for example, the Lipari Islands in the Mediterranean.

obsidian

snowflake obsidian

banded obsidian

pumice
5cm
pitchstone

Ash flow tuffs

If a volcanic eruption is particularly explosive, much extruded material becomes fragmented and is blown up into the air by the explosion. This material will ultimately settle back on to the land surface and, in time, will become consolidated to form beds and layers of pyroclastic rocks.

During the 1902 eruption of Mount Pelée on the island of Martinique in the West Indies, a cloud of extremely hot gas, 700 to 1000°C (1300 to 1850°F), which buoyed up hot fragmentary volcanic material, was seen to flow down the side of the volcano at speeds approaching 160 kph (100 mph). Such clouds have subsequently been observed during many later volcanic eruptions, including the May 1980 eruption of Mount St Helens in Washington State. These gas clouds are called *nuée ardentes* (glowing avalanches). The dense, basal part of these clouds, which contains the solid fragments, hugs the ground surface and flows over it, leaving behind a deposit of loose fragments. These consist of rock fragments, picked up from the ground surface over which the *nuée ardente* flowed and derived from the lavas associated with the eruption, crystals, often of plagioclase feldspar, sanidine, quartz and, less commonly, coloured minerals, such as biotite, hornblende, and pyroxene as well as small glass fragments, often triangular in shape, called glass shards, and lumps of pumice of varying sizes. Such a deposit, as it was formed from an ash flow, is termed an **ash flow tuff**.

These commonly take the form of extensive flat sheets. In a thick sheet the layers nearer the bottom are subjected to pressure from the layers above, so that the shards and pumice, which are still plastic due to heat retention, become squeezed together. The rock develops a streaky appearance and may be referred to as a **flattened tuff**. If the shards become fused together the rock is known as a **welded ash flow tuff**. Squashed, flattened pumice fragments are called **fiammé**. Ash-flow tuffs are also called **ignimbrites**, but some geologists apply this term only to welded ash flow tuffs.

Ash flow tuffs are generally white or grey, although they can be pink and sometimes black, but they all weather to a pale grey colour. The fiammé are easily seen because they are black and glassy, a dark-green colour (due to alteration to chlorite), or a pale grey-white colour (due to conversion to silica). Phenocrysts may be recognized both within the fiammé and the fine groundmass.

Occurrence

Ash flow tuffs are generally confined to sheet-like bodies and are often associated with extrusive acid rocks. Owing to the abundance of silica in these rocks, they are extremely hard and must be extracted with care. The sheets often form angular, craggy outcrops which are frequently well jointed. Ash flow tuffs are found in New South Wales in Australia; parts of Scotland; the English Lake District; Snowdonia in North Wales, and the Cascades range, western America.

ash flow tuffs of different ages
displaying the distinctive streaky appearance

Altered granitic rocks

Few granitic rocks remain completely unchanged following their crystallization from a magma to a solid rock. Usually, these rocks have been variably modified by the activity of hot, pegmatite-producing fluids and gases left behind after the bulk of the magma has solidified. Quite often, these gases and fluids are enriched in certain elements, such as boron, and gases, such as fluorine. Thus, after the final solidification of the magma, but while the rock is still hot, at around 500°C (900°F), these remaining gases and liquids are released and escape through joints and other associated fractures produced by the intrusion of the magma. The action of these fluids and gases can produce changes in the minerals which make up the granitic rocks and cause them to be altered. Hot gases may result in **tourmaline-rich** rocks or cause a process known as **greisening**, while hot fluids cause **kaolinization**.

Greisening

A granitic rock which has undergone greisening is essentially composed of muscovite mica and quartz. Greisens are usually found at the margins of granitic rock bodies, where they are in contact with the country rock. The alteration is restricted to certain zones which often grade into unaltered granitic rock. It seems that the feldspars, essential minerals in the granitic rocks, have become altered to clumps of muscovite, or to a rare mica called **zinnwaldite**, which contains lithium and fluorine. Other minerals often present include topaz and fluorite. Due to the abundance of muscovite and quartz, greisens are leucocratic, sparkling, medium- to fine-grained rocks, although the presence of fluorite can produce purple patches. Topaz, a hard, transparent mineral, is a common accessory and may be more abundant than the muscovite, sometimes comprising up to 90 per cent of the rock. Tourmaline, a boron-rich mineral, may also be present.

Occurrence

Greisens are found, for example, in the Erzgebirge district in Germany, around the granites of Cornwall in south-west England, in the Mourne Mountains of Northern Ireland, and around the granites in Nigeria.

Greisens can sometimes be found as vein-like bodies and also as relatively large masses located at the margins of intrusive granitic bodies, such as the greisen around the Skiddaw granite at Grainsgill in the English Lake District. Topaz and tourmaline are absent from these, and it is possible that some alteration by hot fluids may also have occurred.

Mineralogy

Accessory minerals include tourmaline, apatite, fluorite, cassiterite, rutile, and wolframite.

RUTILE These crystals are prismatic, or slender and needle-like. Rutile crystals usually show one good cleavage and are often reddish

greisen
5cm
orthoclase
quartz
plagioclase
rutile
wolframite and quartz
topaz
fluorite
muscovite
cassiterite

brown, but can be yellowish red or black. They give a pale-brown streak and frequently display a brilliant lustre. Rutile has a hardness of 6½.
WOLFRAMITE These are well-formed, grey-black crystals, often flattened, and showing one excellent cleavage. Single crystals are heavy and possess a dull metallic lustre. Wolframite gives a brownish-black streak and, with a hardness of 4, is easily scratched with a penknife.

Tourmalinization

Tourmaline is a frequent constituent of granitic rocks but, when it is abundant, most of it would have developed after the crystallization of the rock by the process called tourmalinization. Boron-rich gases and, to a certain extent, fluids are responsible for this process. In this case brown tourmaline is produced in place of the essential mineral, biotite mica. As the process progresses, feldspar, also an essential mineral, is replaced, usually by quartz. There are two stages to tourmalinization.
STAGE ONE This is more common and is represented by a rock containing quartz, feldspar, and tourmaline. The quartz shows its usual appearance while the feldspar is brick red and often partly altered. Two types of tourmaline are found: a yellow-green form, which was present in the granitic rock before alteration occurred, and later black to greenish-black tourmaline which occurs as delicate needles in radiating clusters growing from the edges of feldspar crystals and embedded in quartz. These clusters are called tourmaline suns and, although the tourmaline can be easily recognized, it is not always possible to see the clustered arrangement with the naked eye. The coarse-grained rock is called **luxullianite**, and has a striking appearance of dark-green and brick-red patches; the rock itself is sometimes rather crumbly.
STAGE TWO This involves the replacement of feldspar by quartz and tourmaline, which leads to a quartz-tourmaline rock, sometimes called either **roche rock** or **quartz schorl**. In this distinctive black-and-white speckled rock, tourmaline is usually found as discrete stout crystals, rather than in radiating clusters. Examples of such rocks are commonly associated with the Cornish granites.

Kaolinization

This is the process by which the feldspars in granitic rocks are either partially or completely altered to very fine flaky minerals, called clay minerals (of which **kaolin** is one), and very fine, colourless mica called **sericite**. The chief cause of kaolinization is probably extremely hot water which attacks the feldspars. If it has been completely kaolinized, the granitic rock is leucocratic, very soft and crumbly, with only the quartz surviving unaltered. These rocks are mined for the kaolin, or **china clay**, which is of great importance in the pottery industry.

If kaolinization is only partially complete, a **chinastone** is formed. Apart from quartz, partially kaolinized feldspar and some muscovite and sericite are found. Other minerals present include tourmaline, pyrites, topaz, and fluorite. The granitic rocks of Cornwall have suffered kaolinization, and these rocks have been extensively mined for kaolin.

kaolinized granite
luxullianite
5cm
chinastone
tourmalinized granite
tourmaline
in quartz
orthoclase
tourmaline
topaz
quartz-schorl rock
muscovite

Syenites

Syenites are medium- to coarse-grained, intermediate, intrusive igneous rocks. They are usually leucocratic, but can be mesocratic, being either white or shades of grey, pink, and red. The most common minerals are the feldspars (up to 90 per cent), of which the alkali feldspars are always most abundant. Strictly, syenites contain both alkali feldspar (orthoclase and microcline) and plagioclase feldspar. If only alkali feldspar is present, the rock is called an alkali syenite. The amount of quartz is variable. Syenites with up to 10 per cent quartz are called quartz syenites but, more often, syenites have only 2 to 3 per cent visible quartz or may have none at all. Often the feldspars are complexly intergrown to form perthite but often this cannot be recognized in a hand specimen.

If quartz is absent, minerals called **feldspathoids** may be present. Dark minerals are quite frequent, the most common being biotite mica and hornblende. Biotite is found as dark flakes within the rock, the excellent cleavage planes give a glassy or vitreous lustre. Hornblende and other amphiboles may be found as small, dark, prismatic crystals. Other less common dark minerals include pyroxenes and olivine.

Syenites are entirely crystalline, usually with even grain size, but can be porphyritic. Any phenocrysts are of elongate or tabular, white, cream, or pink feldspars which may show some flow alignment.

Occurrence

Syenites are not common. The finer-grained varieties (microsyenites) are found in dyke- and sill-like bodies, while coarser-grained syenites are restricted to intrusive bodies often associated with granitic rocks. Although syenites are relatively uncommon, one particular type, **larvikite**, from Larvik in Norway, is commonly used as a facing stone on buildings. It is a coarse-grained, blue-grey rock and, when polished, the feldspars show a beautiful blue play of light called the **schiller effect**. Examples of syenites include the Plauen area around Dresden in Germany; the Singida district in Tanzania; the type area from Syene (Aswan) in Egypt; Shonkin Sag in the Bearpaw Mountains, Montana; Biella in Italy; Glenelg, Inverness-shire, and the Ben Loyal and Loch Ailsh complexes in Sutherland, Scotland.

Mineralogy

The essential minerals are feldspars, biotite, and hornblende. Quartz and feldspathoids (nepheline and sodalite) may be essential in some syenites, but never occur together. Apatite, zircon, pyrite, sphene, magnetite, pyroxenes, and olivine are common accessories.

NEPHELINE White, often hexagonal, glassy crystals, with a hardness of 6, no cleavage, and a conchoidal fracture. It can be confused with quartz, but nepheline can just be scratched with a penknife.

SODALITE Its physical properties are very similar to those of nepheline, but it is a distinctive bright blue.

syenite
nepheline syenite
larvikite
rhomb porphyry"
(porphyritic microsyenite)
perthite
amazon stone (microcline)
plagioclase
quartz
nepheline
sodalite
biotite
hornblende
augite
5cm

Trachytes

Trachytes are fine-grained, intermediate, igneous rocks and are the extrusive equivalents of the syenites. They have the same relative mineral composition as the syenites. Trachytes are leucocratic, usually being white, pink, or a pale creamy yellow, but can be mesocratic. Very commonly, they are porphyritic; the phenocrysts are usually feldspar but biotite and hornblende may also occur. Trachytes have a rough feel owing to the many tiny bubbles or vesicles. The name trachyte is derived from the Greek *trachys* meaning rough.

The phenocrysts in trachytes, which may sometimes be seen in hand specimen, often show an alignment, because of flow movement in the original magma. The abundant, lath-like feldspar crystals have a crude parallel arrangement, giving rise to a **trachytic texture**.

Alkali feldspar is the dominant mineral, of which the varieties called sanidine and anorthoclase occur, but they are very difficult to distinguish from other alkali feldspars in hand specimen. Plagioclase feldspar and quartz may be present in small amounts or may be absent. Small, well-formed flakes of biotite are present, together with hornblende, while some trachytes have unusual sodium-rich amphiboles which may develop a blue colour. Pyroxene and olivine may be present but are usually never abundant. A variety of feldspathoids can be common in trachytes which have no quartz; such rocks are called **phonolites**. The fine pale groundmass is predominantly composed of alkali feldspar.

Occurrence

Trachytes occur as dykes and plugs, short, thick lava flows, or domes. The original lava was viscous and did not flow easily, so that trachyte lava flows do not achieve any large lateral extent. They exert no real influence on the landscape but the dykes and plugs may form upstanding areas because they are harder than the surrounding country rocks. Examples are obtained from the Cripple Creek district, Teller County, central Colorado; the Drachenfels area in Germany; the Auvergne district of south-central France (domites of Puy de Dome); Phlegraean Fields in Italy; the Banks peninsula, New Zealand, and Tuolumne County, east-central California. Examples in Scotland include those from Breigh a Choire Mhoir on the Isle of Mull and the Eildon Hills near Melrose in southern Scotland.

Mineralogy

The essential and accessory minerals of trachytes are the same as those of syenites (*see* page 84), but only those minerals forming the phenocrysts in trachytes can be recognized in hand specimen. The different feldspathoids will not be described, because they are difficult to identify in hand specimen.

biotite
plagioclase
hornblende
5cm
quartz
biotite trachyte
(domite)
trachyte
hornblende
trachyte
porphyritic
trachyte
sanidine trachyte
trachyte

Diorites

Diorites are medium- to coarse-grained, intermediate, intrusive igneous rocks. They are generally darker in colour than the acid rocks and syenites, although their colour is variable, ranging from mesocratic to melanocratic. Diorites are distinctive black or dark-green and white speckled rocks in hand specimen. The light- and dark-coloured minerals are segregated to form irregular 'clots' or even distinct layers.

Mineralogy

The mineralogy of diorites is relatively straightforward. They can possess small amounts of quartz (2 to 3 per cent) when they are called quartz diorites. Quite commonly these also contain biotite mica, the most abundant of the coloured minerals. Quartz diorites are somewhat lighter in colour than the diorites.

The dominant light-coloured mineral in diorites is plagioclase feldspar, especially the plagioclases oligoclase and andesine, but it is not possible to distinguish them in a hand specimen although they often occur as quite well-formed crystals. The absence or near-absence of quartz enables diorites to be distinguished from granitic rocks, while the dominance of plagioclase feldspar allows a distinction to be made from syenites, in which alkali feldspar is dominant.

Alkali feldspar, either as orthoclase or microcline, can be present but is generally scarce. Dark-coloured minerals are more common in diorites than in the granitic rocks and the syenites. The characteristic dark-coloured mineral in diorites is the amphibole hornblende which is found as fairly well-formed, dark green, brown, or black prismatic crystals. Sometimes, in melanocratic diorites, hornblende can replace plagioclase feldspar as the dominant mineral. Biotite mica, found as small flakes is also present, though it is less abundant than hornblende. It is more common in quartz diorites.

Other dark-coloured minerals present are pyroxene and olivine but they only rarely become important in diorites. The typical mineral composition of a diorite is plagioclase feldspar, hornblende, and a little biotite.

Texture

Most diorites possess crystals of roughly equal size. The occurrence of the light- and dark-coloured minerals in clots may make diorites appear much coarser grained than they really are. They may be porphyritic, the phenocrysts commonly being plagioclase feldspar, hornblende, or biotite. Poikilitic texture is often developed in diorites, in which hornblende partially or wholly encloses plagioclase feldspar crystals. Textures are variable and can even vary within a hand specimen. Xenoliths are often present which, if not too heavily altered and recrystallized, may reflect the type of country rock into which the diorites were intruded.

markfieldite
diorite
diorite with
hornblende phenocrysts
porphyritic
microdiorite
diorite with
quartz and biotite
5cm
microdiorite
showing weathering
'mela' diorite

Occurrence

Diorites are relatively uncommon and seldom form large independent masses. They are more usually found as small intrusive bodies and plugs, less frequently in large dykes and sills. They are often located on the margins of large granodiorite and gabbro intrusions.

Diorites and dioritic varieties are recorded from the Garabal Hill complex in the Glen Fyne area, and in Argyllshire, Scotland; Stavanger and Trondheim, in Norway; Newry area in Northern Ireland; small intrusive bodies in the Coast Range batholith in south-east Alaska; south-east Jersey and northern Guernsey, Channel Islands, and Pfaffenreuth, Passau, Bavaria.

Outcrops are usually somewhat weathered, thus surface specimens will not show a typical appearance.

MICRODIORITES These rocks have the same mineral composition as diorites but are finer grained. Most are porphyritic, with phenocrysts of plagioclase feldspar, hornblende, or biotite. The surrounding groundmass is much finer and individual crystals are often not visible to the naked eye. Such rocks are most commonly found as sills and dykes, such as those microdiorites associated with the Lausitz and Riesengebirge granodiorites in Germany, and a microdiorite dyke cuts the Chosica pluton in Peru. Several examples of **porphyritic microdiorites** are recorded from Scotland, where they are termed porphyrites; for example, the Glencoe-Ben Nevis area, the Cheviot Hills, and those associated with the granodiorites in the Galloway area of south-west Scotland. Porphyritic, quartz-bearing microdiorites, called **markfieldites** (*see* page 89), are recorded from Markfield in the Charnwood Forest area in Leicestershire.

Essential minerals

The essential mineralogy of diorites and microdiorites consists of plagioclase feldspar, hornblende, and biotite.

PLAGIOCLASE FELDSPAR Crystals can be prismatic or tabular but can be found in massive aggregates. They are generally white or grey, but can acquire a reddish tinge. Plagioclases give a white streak, and show two good cleavages – the cleavage planes having a vitreous lustre. They can just be scratched with a penknife, having a hardness of 6. The white colour of plagioclase feldspars often distinguishes them from the pink alkali feldspars.

HORNBLENDE This is often found as dark green, brown, or black prismatic crystals showing two good cleavages which intersect at approximately 120° and usually show a vitreous lustre. Well-formed crystals may have hexagonal outlines and can be scratched with a penknife (hardness 5 to 6).

Accessory minerals

Common accessory minerals include pyroxene and olivine (*see* page 100), quartz, orthoclase, apatite, zircon, sphene, pyrite, chalcopyrite and magnetite.

plagioclase

hornblende

biotite

diorite

5cm

quartz

apatite

zircon

augite

sphene

granular aggregate of magnetite

chalcopyrite

pyrite

Andesite

Andesites are fine-grained, intermediate, entirely crystalline igneous rocks and are the extrusive equivalents of diorites. They have the same mineral composition as diorites. Andesites are generally dark, ranging from mesocratic to melanocratic. In hand specimen, they may vary in colour from black to green, dark red, brown, purple, and grey and often show a spotting, due to the presence of pale phenocrysts surrounded by a much finer-grained dark groundmass. Rarely they are leucocratic. Although the minerals are essentially the same as those in diorites, only those minerals found as phenocrysts can be identified. Crystals in the groundmass are much too small to be identified even using a hand lens.

Mineralogy

The most important constituent of andesites is plagioclase feldspar, which is found as white or grey phenocrysts and as tiny lath-like crystals in the groundmass. Plagioclase may make up as much as 60 to 70 per cent of andesites. The common plagioclases present are andesine and oligoclase. Small crystals of alkali feldspar are present in the fine groundmass, but are generally rare. Some quartz is present in the groundmass but is not abundant and is never found as phenocrysts.

Dark minerals, especially pyroxenes, are common in andesites. Augite is present both in the groundmass and as phenocrysts as is hypersthene, although enstatite is occasionally recorded. It is difficult to differentiate accurately between these in hand specimen.

Other coloured minerals present are hornblende, biotite, and olivine. Olivine is never abundant, but can be found forming phenocrysts in more basic andesites which are darker in colour. These have less silica than a 'typical' andesite. Those bearing olivine phenocrysts are termed olivine andesites. Pyroxene andesites are also recorded and have abundant plagioclase feldspar phenocrysts, as well as common pyroxene phenocrysts. In contrast with their coarse-grained equivalents, the diorites, hornblende and biotite are comparatively rare in andesites, but do sometimes form phenocrysts.

Originally, the phenocrysts seen in andesites crystallized out within the magma before it reached the surface, rather like the pips in raspberry jam. Often, these phenocrysts reacted with the hot, molten magma, causing their outlines to become corroded and smoothed off. This feature may be seen using a hand lens.

Texture

The dominant texture displayed by andesites is porphyritic, although the groundmass may show a trachytic texture. As andesites are extrusive rocks, they have been subjected to rapid cooling so that glass is occasionally present within the fine groundmass and, in a few cases, it may be completely glassy. Quite often the groundmass is composed entirely of tiny crystals. Andesites can also develop tiny gas holes or vesicles, which

altered andesite

vesicular andesite

5cm

andesite

andesite
with hornblende
phenocrysts

porphyritic andesite

andesite
with hornblende
phenocrysts

andesite

may be infilled later by minerals. Andesites are susceptible to alteration, especially to a type associated with later igneous activity after the andesite had solidified. The rocks develop a dull green colour, due to the progressive replacement of the minerals by the green minerals chlorite and epidote, and calcite.

Occurrence

Andesites are common volcanic rocks, second only to basalts in abundance. They are often found as lava flows and less commonly as dykes. They weather fairly quickly, so patience is required when searching for fresh specimens. The lava flows do not exert any widespread effect on the landscape, apart from producing upland areas when the flows are particularly extensive.

Examples include andesites from the Cascades range, western North America (Mount Rainier, Mount Hood); many of the volcanoes of the Andes, South America (the volcano Paricutin); and the Aleutian, Javanese, Japanese, and south-western Pacific islands; the Aegean Islands in the Mediterranean; the West Indies; Brokeoff Peak, Lassen Volcanic National Park, western USA; western Sidlaw Hills, Perthshire in Scotland; Vatnajokull, Iceland, and the English Lake District.

Essential minerals

These are plagioclase feldspar and hornblende, biotite, pyroxenes, and olivine.

Accessory minerals

These include apatite, zircon, magnetite and pyrite, although they are not recognizable in hand specimen.

andesite

plagioclase

hornblende

biotite

5cm
quartz

augite

granular olivine
aggregate

chlorite

Gabbros

Gabbros are coarse- to medium-grained, basic, entirely crystalline intrusive igneous rocks. There are several different kinds, depending upon the differing combinations and percentage abundance of certain minerals, so that the term gabbroic rocks may be more accurate. Gabbroic rocks generally are dark, ranging from mesocratic to melanocratic, and occasionally may be described as hypermelanic. They often possess a speckled appearance.

Mineralogy

The only common light-coloured mineral is plagioclase feldspar, specifically, the varieties labradorite and bytownite, but their presence is rather variable. In some cases they may constitute 60 per cent of the rock.

Several dark-coloured minerals are found within gabbroic rocks, the most important being the pyroxenes, including hypersthene and, less commonly, bronzite, although it can be difficult to tell the two apart in a hand specimen.

Augite is commonly present and, in gabbroic rocks enriched with calcium, titanaugite occurs. It is very difficult to differentiate between these minerals accurately in hand specimens.

Another common coloured mineral is olivine. It is not usually as abundant as pyroxenes but it is often present as rounded crystals. These can be distinguished by their pale-green colour, although this may be masked by the darker greens and blacks of the pyroxenes.

Other coloured minerals which are found in gabbroic rocks are hornblende and biotite, although they are never as common as the coloured minerals mentioned earlier. A deep-brown amphibole, kaersutite, which is related to hornblende, is often found in calcium-rich gabbroic rocks. Very rarely, quartz and alkali feldspar may occur, but it is doubtful whether these minerals could be recognized in hand specimen. As a general rule, quartz and olivine are never found together in the same igneous rock.

Types of gabbroic rocks

Strictly, a gabbro consists only of plagioclase feldspar (labradorite) and augite. If some hypersthene is present, the rock is called a hypersthene gabbro; where only plagioclase and hypersthene are present, the rock becomes a **norite**; where some olivine is found it is termed an **olivine gabbro**. Sometimes the rock may be simply composed of olivine and plagioclase feldspar (labradorite) when it is called a **troctolite**. If augite is associated with the plagioclase feldspar, bytownite, it is an **eucrite**.

Texture

Gabbroic rocks are coarse grained, and entirely crystalline. The crystals themselves all interlock and rarely show well-developed shapes. Finer-grained varieties are called microgabbros. Nevertheless, gabbroic rocks

troctolite
gabbro
norite
eucrite
quartz gabbro
'mela' gabbro
gabbro
5cm

show a range of textures. Poikilitic texture is common where a large crystal is seen either to enclose partially or completely several smaller crystals. Frequently in gabbroic rocks, large crystals of augite wholly or partially enclose small lath-like plagioclase crystals. This variety of poikilitic texture is given the name **ophitic texture** and is only applied when augite encloses plagioclase feldspar. It is also extremely common in finer-grained basic rocks, such as dolerites (*see* page 102).

Porphyritic texture is rare in gabbroic rocks. Troctolites may seem porphyritic but this is caused by dark olivine crystals standing out against the aggregate of grey plagioclase crystals, hence its popular name 'troutstone'; in fact, the plagioclase crystals are approximately the same size as the olivines.

Banding

A feature often seen in gabbroic rocks is **banding** or **layering** of one type of mineral. When magmas, especially intrusive ones, cool and slowly solidify, minerals crystallize in a particular order.

In a slowly cooling basic magma body, olivine crystallizes first; as the magma has a low viscosity, these fairly dense crystals very gradually sink to the bottom of the magma body and slowly accumulate; the pyroxene crystals then do the same. The plagioclase crystals are not as dense so they float rather than sink. Bands rich in chromite are often associated with the olivine and pyroxene bands and these are mined for the chromite. The layers range in thickness from a fraction of a centimetre up to 30 cm (1 ft) and are repeated in a rhythmic fashion, possibly due to the influx of fresh magma. These layered sequences often display features, such as cross-bedding, more commonly associated with sedimentary rocks. An excellent example of a layered body is that of the Palisades Sill, New York State, located along the Hudson River, in the USA.

Occurrence

Gabbroic rocks are often weathered and the exposed surfaces are usually a reddy brown, rusty colour. This colour is related to the fairly high iron content of many such rocks. When fresh, these rocks are hard and difficult to break and extract.

Gabbroic rocks are found in a variety of intrusive forms, such as laccoliths and lopoliths (*see* page 23), which are often very large. The Duluth lopolith in Minnesota, for example, is calculated to contain around 200,000 cubic kilometres (48,000 cubic miles) of gabbroic rocks. Such bodies have a profound effect on the landscape and usually form large upstanding areas.

Gabbroic rocks are common worldwide. Examples include the layered complexes of Skaergaard in east Greenland; the Bushveld (Transvaal) in South Africa; the Stillwater complex in Montana; the Freetown complex in Sierra Leone, and, on a smaller scale, the St Peter Port gabbro, Guernsey, Channel Islands. Other examples are Carlingford, County Louth, southern Ireland; Ardnamurchan, western Scotland; the Lizard complex, Cornwall; south-west England; Sudbury area of central

layered gabbro
5cm
olivine gabbro
chromite bands in
weathered gabbro
nodular chromite
hypersthene gabbro
hypersthene gabbro
gabbro

Ontario, Canada; the San Marcos gabbros of southern California, and the Wichita mountains of Oklahoma.

Essential minerals

These are plagioclase feldspar (labradorite or bytownite), pyroxene, and olivine with lesser amounts of hornblende and biotite.

PLAGIOCLASE FELDSPAR The common variety is labradorite. Although plagioclases are usually almost impossible to differentiate in hand specimen, occasionally, labradorite may be identified by the spectacular play of blues and greens displayed on cleavage surfaces, which is a type of schiller effect.

PYROXENE Hypersthene is found as irregular dark brown-black crystals in gabbroic rocks. The variety bronzite has a bronzy yellow colour. Hypersthene has a white streak and a vitreous lustre. It may be possible to recognize two cleavages, which are at right-angles to each other. Crystals can only be scratched with difficulty, having a hardness of 6. Well-formed crystals of augite have square or eight-sided cross-sections and are usually dark green-black. The other physical characteristics (hardness, cleavage, lustre) of augite are very similar to hypersthene, making these two pyroxenes very difficult to distinguish in hand specimen.

OLIVINE Well-formed crystals are rare. Olivine is usually olive green in colour but can be white, red, brown, or black. It does not show a good cleavage. It has a white streak and a vitreous lustre. A penknife will only scratch olivine with difficulty (hardness 6½). The most distinctive feature of olivine is its colour. It is easily altered by hot fluids into serpentine.

SERPENTINE This is a group name, encompassing the minerals chrysotile and antigorite. Serpentine occurs either in structureless masses or, more commonly, in fibrous form. Usually it is seen in shades of green but it can also be browny yellow and grey. Hardness is variable (4 to 6) so it can often be scratched with a penknife. Massive aggregates have a greasy lustre, while fibrous forms have a silky lustre.

Accessory minerals

Accessory minerals in gabbroic rocks include quartz, alkali feldspar, apatite, zircon, pyrite, chalcopyrite, sphene, spinel, and chromite.

SPINEL Well-formed crystals are eight sided (octahedral) but usually they are massive. Crystals may be red, blue, green, brown, black, or colourless, and they can be transparent. They have a vitreous lustre, no cleavage, and a white streak. A hardness of 8 means spinel is almost impossible to scratch. The crystal shape and hardness are diagnostic.

CHROMITE Chromite is usually found as massive black-brown aggregates. The crystals have no cleavage, show a bright metallic lustre, and yield a dark-brown streak. Chromite aggregates are fairly heavy and crystals can be scratched with a penknife (hardness 5½). These may be confused with magnetite but the brown streak of chromite is diagnostic.

labradorite
(plagioclase)
plagioclase
hypersthene
bronzite
augite
olivine
aggregate
hornblende
serpentine
gabbro
biotite
spinel
1cm
pyrite
chalcopyrite
apatite
5cm
chromite

Dolerite (diabase)

Dolerite is a medium-grained, basic, entirely crystalline, shallow intrusive igneous rock. It is finer in grain size than the gabbroic rocks, but is coarser grained than basalts. In America, dolerite is called diabase while, in Europe, a diabase is a metamorphosed dolerite.

Mineralogy

Dolerites have the same relative mineral composition as the gabbroic rocks, with essential minerals of plagioclase feldspar, pyroxene, and olivine. They range from mesocratic to melanocratic, generally having a speckled black-and-white appearance, due to the random distribution of light plagioclase and dark pyroxene and olivine. The grain size is variable, often approaching the grain size of gabbroic rocks.

Dolerites occasionally have a little quartz present (for example the Whin Sill dolerite) and, if cooled quickly, some brown glass. Neither the quartz nor the glass would be identifiable in hand specimen, however. Some hornblende and biotite may be present, as well as a little alkali feldspar and some iron ore.

Texture

The rocks appear granular in hand specimen and are only rarely porphyritic. Ophitic texture is particularly common, and may be recognized with a hand lens. Gas holes or vesicles may be seen, possibly infilled by minerals such as calcite and zeolites, forming amygdales.

Occurrence

Dolerites are susceptible to weathering and alteration. Surface outcrops often develop a reddy brown coloration, with plagioclase crystals standing out in relief. Such surfaces have a somewhat pitted appearance, although the outcrops themselves are rather smooth and rounded with few angular edges. Due to their rounded form and weathering, fresh dolerite specimens may be difficult to obtain but, with patience, adequate specimens can be collected. The rock is often well jointed.

Dolerites are usually found as dykes and sills, often at the edges of large gabbroic bodies. Hundreds of these dykes may be found in any one area in a dyke swarm (*see* page 23). Depending on the structure of an area, dolerites may exert some limited influence on the landscape, because they are relatively hard rocks which can form upstanding masses. They are common worldwide. Examples include the Karroo dolerites from South Africa; the Mount Wellington Sill, Tasmania; the Palisades Sill, New York State; and dolerites on the Isles of Mull and Arran, off western Scotland. Examples from England and Wales include those from the English Midlands (Rowley Regis, near Birmingham); the Clee Hills in Shropshire; the Whin Sill; localities in Derbyshire, and from Snowdonia in North Wales.

dolerite
diabase
5cm
olivine dolerite
dolerite showing weathered skin
olivine dolerite

Basalts

Basalts are fine-grained, basic, entirely crystalline, extrusive igneous rocks. The majority of basalts are dark, usually melanocratic. The common colours developed are black or greyish black, but some basalts may develop red and green tinges. In general, the individual crystals within basalts are not visible with the naked eye or through a hand lens. Occasionally, relatively large phenocrysts may be present.

Mineralogy

The mineral composition of a basalt is essentially the same as that of a gabbro, although it may be difficult to determine this from examination of a hand specimen. The characteristic mineralogy of basalts is plagioclase feldspar (labradorite, bytownite, and anorthite), pyroxene, iron ore, and olivine, although some basalts may lack olivine. Depending on the presence, absence, and abundance of particular minerals, basalts can be split up into different groups, although these divisions cannot be accurately applied to hand specimens.

The common plagioclase feldspar in basalts is either labradorite or bytownite which may occur as phenocrysts and are also present in the very fine-grained groundmass. As in other basic rocks already described, pyroxene is the dominant coloured mineral and mostly accounts for the dark coloration of basalts. Augite is the most abundant pyroxene in basalts, although pigeonite and titanaugite also occur. Augite often occurs as phenocrysts while pigeonite and titanaugite are only rarely found as phenocrysts. Hypersthene is also found in basalts and may occasionally form phenocrysts.

Olivine is found in most basalts, only rarely achieving dominance over the pyroxenes. It often forms phenocrysts and may also be found in the groundmass. Hornblende and biotite are fairly rare in basalts and they are unlikely to be recognized in hand specimens.

Iron ores are common in basalts, especially magnetite and ilmenite, and they contribute to the overall melanocratic appearance of basalts but rarely occur as phenocrysts. Quartz and alkali feldspar are very rare in basalts although, occasionally, isolated quartz crystals can be recognized under a microscope.

Basalts can be classified in different ways but most schemes require microscopic examination of the rock to achieve any accuracy and, in hand specimen, the best we can do is to identify the rock as a basalt and try to recognize any visible phenocrysts.

Features and textures

Rock fragments or xenoliths are relatively common and provide useful information about the type of rocks through which the basaltic magma passed on its way to the Earth's surface. Gas holes or vesicles occur frequently and may remain open. Where they have been infilled the rock is referred to as an amygdaloidal basalt (*see* page 107). White, fibrous,

hypersthene
olivine aggregate
augite
magnetite aggregate
labradorite (plagioclase)
basalt
5cm

moderately soft zeolites often form the infilling, as do calcite and chlorite, in addition to varieties of silica called agate and chalcedony. Agate amygdales often show concentric banding developed on a fine scale. Although not visible in hand specimen, some basalts contain brown glass in the groundmass.

The textures developed in basalts are variable and cannot all be seen in hand specimen. Porphyritic basalts are relatively common, however, with the groundmass often being trachytic. Occasionally, several phenocrysts of the same mineral seem to have become clustered together, forming a clot which, at first sight, may seem to be one large crystal. This **glomeroporphyritic texture** is not restricted to basalts and can be developed in any porphyritic rock. Take care not to confuse porphyritic basalts with amygdaloidal basalts because they can often appear rather similar in hand specimen.

In general, most basalts appear granular in hand specimen.

Lava flows

Basalts are the most abundant extrusive igneous rocks, and their total surface area exceeds that of all the other extrusive igneous rocks combined. They most commonly occur as lava flows, many of which are very extensive. Basalt magmas are very hot and fluid, and these factors, coupled with the immense volumes of lava which are often erupted, enable basalt lavas to flow over and cover many square kilometres. For instance, basalts which are found on the Columbia River-Snake River plateau of Oregon, Washington and Idaho in the western United States, cover more than 500 000 square kilometres (about 200 000 square miles). The temperature of basalt magma is approximately 1100°C (2000°F) when it is erupted or it can be even hotter. As it flows, it cools and becomes more viscous. The surface of the lava in contact with the air cools more quickly than the interior of the flow and a crust forms on top. This can be rather thin, forming a skin over the lava, or quite thick, developing a broken rubbly character. The thin, skin-like crust is particularly well developed on the basalt flows erupted on the Hawaiian islands. As the skin is often wrinkled and puckered, it takes on the appearance of strands of coiled rope, so it is called ropy lava (*see* page 109), or, in Polynesian, pahoehoe. The rubbly top to basalt lavas is more common and is called blocky or aa lava. These surface features are seen to be developed in recent (over the last 10 000 years) and in ancient flows. Recent basalt flows are often associated with other products of volcanic eruptions such as ash deposits.

JOINTING As an igneous rock cools, it contracts, causing the newly formed rock to crack and fracture. Such fractures are called joints. In thick basalt deposits, such as sills, the jointing takes on a regular structure and forms four-, five-, six-, seven-, and eight-sided columns of rock, approximately 60 cm (2 ft) across. This is called columnar jointing and the best-known examples are found at the Giant's Causeway in County Antrim, Northern Ireland and Fingal's Cave on the Isle of Staffa.

PILLOW LAVA Basalt lavas may also be extruded completely

zeolite
calcite
vesicular basalt
5cm
varieties of amygdaloidal basalt

under water to form a pillow lava (*see* page 21). The outermost part of a pillow, which cooled very quickly, is very fine grained and may even be glassy. Vesicles are also present, often arranged in concentric bands within the pillows. Pillow lavas can be recognized in sequences of rocks over 3000 million years old.

Occurrence

Basalts are relatively susceptible to weathering because olivine is very easily altered to serpentine and other products while the pyroxenes can break down to form chlorite. These rocks are often subjected to **spheroidal weathering**. Here, the surface of the basalt, which may often acquire a reddish tinge, breaks away in successive layers, rather like the skin of an onion, so it is popularly called onion-skin weathering, and gives the basalt a rounded appearance. It is particularly well-developed in jointed basalts. Nevertheless, basalts are relatively hard rocks and can form upstanding areas. The large sheet-like flows of basalt, as found in the Columbia River area, often form large plateaus and are known as plateau lavas. Basalts are also commonly found in dykes and sills.

Basalts erupted under water also suffer alteration to rocks called **spilites**. The mineral changes are often complex and will not be discussed in detail, but the following do occur: labradorite is replaced by another variety of plagioclase called albite; and the common coloured minerals, pyroxene and olivine, are replaced by chlorite which gives spilites a greenish colour, although some augite may remain.

The extensive jointing, coupled with weathering, enables basalt specimens to be extracted fairly easily, although exposures showing good examples of columnar jointing and pillow lavas, should be admired and *never* hammered.

As basalts are so common worldwide, examples abound. The Hawaiian islands are famous for their basalt flows. Examples of large basalt flows include the Deccan area of India; the Karroo plateau of South Africa; the Parana basin of Argentina; the Keweenawan area around Lake Superior, and the New Jersey area, USA; areas in the southern Cascades, USA (Cinder Cone, Mount Shasta, Medicine Lake); the islands of Mull and Skye off western Scotland, and the Midland Valley of Scotland. Spilites are recorded from north Cornwall and south Devon in south-west England, as well as areas in Wales (Pembroke), Anglesey (Newborough Warren), and south western Scotland (Tayvallich peninsula).

The spectacular Giant's Causeway, Antrim, Northern Ireland, displays columnar jointing in basalt.

vesicular basalt
with olivine
phenocrysts
surface of
ropy lava
(pahoehoe lava)
basalt with
pyroxene phenocrysts
basalt with olivine
phenocrysts, vesicles
and gas pipes
spilite
vesicular
porphyritic basalt
basalt
5cm

Ultrabasic rocks

Ultrabasic rocks contain less than 45 per cent silica, by definition. Certain rocks, however, which are proven to be ultrabasic because of other features, such as mineral composition, are found to contain more than 45 per cent silica. Most ultrabasic rocks are dark, coarse grained, intrusive, and entirely crystalline. They may be described as hypermelanic due to a preponderance of dark-coloured minerals, consisting mainly of magnesium and iron. Thus, the rocks are often described as **ultramafic**.

Ultramafic rocks are very variable and include some rather rare rock types. Many are found in layered bodies and some are composed of only one mineral. They have no extrusive equivalents. Ultramafic rocks may be split up into three broad groups: **peridotites**; rocks composed of only one mineral, such as **pyroxenites** and **hornblendites**; and **picrites.**

Peridotites

Peridotites are ultramafic rocks with no plagioclase feldspar and in which olivine is the dominant mineral. In some cases olivine may be the only mineral present and the rocks are known as **dunites** or **olivinites**. Due to the abundance of olivine, dunites are often green in hand specimen and appear granular. In some cases, chromite may be found as an accessory mineral, occurring as black specks, either scattered through the rock or concentrated into diffuse bands of variable thickness.

Peridotites contain other coloured minerals as well as the dominant olivine. Bronzite is common, but never dominant, in **bronzite peridotites** which are themselves the commonest peridotites. Hypersthene also occurs as does augite, and certain peridotites contain both types of pyroxene as well as olivine. Rarely, peridotites containing hornblende are recorded. Other types contain mica (**mica peridotites**), such as an uncommon variety called **phlogopite** which often has a pale yellowy brown colour. Of particular interest is the variety of mica peridotite from the Transvaal in South Africa, called **kimberlite**, which is found in vertical pipe-like bodies; one of its accessory minerals is pure carbon or diamond. **Garnet peridotites** contain some garnet, such as a dark-red variety called pyrope, while **alkali peridotites** possess calcium-rich pyroxenes and amphiboles.

Peridotites and dunites are easily altered due to the abundance of olivine. In some cases, most of the olivine becomes altered to green serpentine (*see* page 100). To a lesser extent, the pyroxenes also suffer some alteration to serpentine. If this type of alteration is widespread throughout the particular peridotite, the rock is called a **serpentinite**. Such rocks are pale grey-green in colour, streaked with red, and often banded or mottled. Good examples in England can be obtained from the Lizard complex in Cornwall.

Ultramafic rocks composed of one mineral

Pyroxenites and hornblendites are examples of these rocks, and are

bronzite

olivine

hypersthene

hornblende

augite

phlogopite

5cm

peridotite

bronzite peridotite

grouped together under the name **perknites**. Their characteristic feature is that they are composed almost entirely of mafic minerals other than olivine, such as pyroxenes and amphiboles. Perknites can be found as small intrusions or, more often, within larger layered intrusions (*see* page 98). The grain size in these rocks varies from medium to coarse and they generally have a granular appearance. Although the terms pyroxenite and hornblendite imply rocks composed entirely of pyroxene or hornblende, accessory minerals are often present.

BRONZITITE This is the most common pyroxenite. It is composed almost entirely of bronzite. In hand specimen, bronzitite often displays a bronzy yellow lustre or sheen. Any accessory minerals are poikilitically enclosed by the bronzite crystals. Such rocks are common and widespread as layers in the gabbro within the Bushveld complex in South Africa and, within this complex, the famous Merensky Reef, which is mined for platinum, is contained within a bronzitite. Commonly associated with bronzitites are bands of pure chromite called **chromitites**.

Accessory minerals within bronzitites are usually other pyroxenes and plagioclase, the pyroxene often being a bright emerald green colour. Other pyroxenes, such as hypersthene and enstatite, may also form the dominant components in pyroxenites, but they are never as common as bronzite.

WEBSTERITE The name Websterite is applied to rocks containing combinations of different types of pyroxenes, such as hypersthene, augite, and diopside after the type locality at Webster, North Carolina, although some geologists prefer self-explanatory terms like **augite-hypersthene pyroxenite**.

OTHER PYROXENITES The accessory minerals present in pyroxenites are used to qualify these rocks, thus **hornblende pyroxenite** contains hornblende, and there are **mica pyroxenites** containing biotite, **spinel pyroxenite**, and **olivine pyroxenite** which is halfway between a pyroxenite and a peridotite.

HORNBLENDITES Hornblendites are very dark rocks, which are almost entirely composed of hornblende and are sometimes associated with diorites, gabbroic intrusions, or other ultramafic rocks. They are often found as small intrusions and dykes. Frequent accessory minerals are pyroxenes and olivine.

GLIMMERITES Other ultramafic rocks which are composed almost entirely of one mineral are those rocks consisting of dark mica, such as biotite or phlogopite. They are generally quite rare and are called **glimmerites** because of their shiny appearance (*see* page 115).

Picrites

The diagnostic feature of picrites (*see* page 115) is the presence of small amounts of plagioclase feldspar (anorthite and bytownite) in addition to olivine, pyroxene, and amphibole. The pale minerals may serve to give these particular rocks a spotted appearance. Picrites are similar to peridotites and grade into them.

dunite
bronzitite
serpentinite
websterite
pyroxenite with
vein of epidote
chromitite
5cm

Anorthosites

These are medium- to coarse-grained, entirely crystalline, intrusive igneous rocks made up of more than 90 per cent plagioclase, in particular, the varieties labradorite, bytownite, or anorthite. Dark-coloured minerals, such as pyroxenes and amphiboles, are frequent accessory minerals. These are certainly not ultramafic rocks, because the preponderance of plagioclase makes them leucocratic, but they are ultrabasic because they contain less than 45 per cent silica. They are often found as separate intrusive bodies which, interestingly, are all usually more than 600 million years old. Younger anorthosites are found as bands and layers within the large layered gabbroic masses already mentioned.

Occurrence

On account of the abundance of minerals such as olivine, pyroxene, and amphibole, ultrabasic and ultramafic rocks are easily weathered and often have a reddy brown surface colour. Fresh specimens may be hard to come by; no special extraction techniques are required, but you should look for protruding angles and corners to avoid having to hammer an outcrop extensively. Only the large layered complexes already mentioned have any significance on the landscape, usually forming large upland areas.

Peridotites are commonly found either as layers within the large layered bodies of the Bushveld in South Africa and the Stillwater complex in Montana, USA, or as pod- and sheet-like bodies in mountain chains such as the Alps, the Urals, and the Appalachians. Specific examples of peridotites and dunites include those from South Island, New Zealand; the Isle of Skye off the west coast of Scotland; near Jefferson in North Carolina, USA; around Vampula in Finland; the Lizard area in Cornwall, south-west England; the Tulameen area in British Columbia, and the Cortland area in New York State.

Pyroxenites are noted from Cecil County in Maryland, USA; the major layered bodies already mentioned; Webster, North Carolina; the Suc Valley of the Ariège district in the Pyrenees, and at Iron Hill in Gunnison County, Colorado.

Hornblendites are recorded from many of the localities mentioned above, in addition to the Garabal Hill area in Argyllshire, Scotland and the Ariège district in the Pyrenees, while glimmerites may be associated with the layered bodies.

Picrites are associated with peridotites and perknites and have similar distributions as well as being noted around Plymouth in Cornwall, Anglesey in North Wales, and Glen Orchy in Scotland.

The large, old anorthosite intrusions may be seen at areas within the Adirondack Mountains in New York State, USA; southern Norway, and areas within Labrador and Quebec in Canada. They can also be found as bands within the large layered complexes already mentioned.

pyroxenite
picrite
hornblendite
glimmerite
5cm

anorthosite

Mudrocks

Mudrocks are the finest-grained detrital sedimentary rocks and the grains are not visible to the naked eye. This group is often divided according to grain size and further subdivided with respect to major mineral components, accessory minerals, sedimentary structures, and diagenesis (*see* page 28).

Clays and claystone

Claystone becomes clay when wet and it also becomes plastic and malleable. It has no well-developed bedding but may show banding due to compositional variation. This group is the very finest-grained of the mudstones and the rocks are composed of grains less than 0.004 mm (0.00015 in). Soft clays are distinctive because of their blocky fracture. The more lithified sediments will be soft and often powdery.

The clay minerals of which they are composed have a sheet-like or layered structure, like the micas and the differing compositions of these layers gives rise to clays with different properties and crystal shape. It is often difficult to identify specific clay types although certain properties, such as the ability to absorb water, distinguishes some of them, as with the **smectites**.

Clay minerals

ILLITE Colourless to yellowish brown, illite is the most common clay mineral in sediments and is formed as an intermediate product during the alteration of feldspars, micas, and other constituents in the soil, particularly in temperate climes. It may also grow after the sediment has been deposited.

KAOLINITE This describes a group of minerals, which are mostly colourless to pale yellow, including dickite, halloysite, nacrite, and kaolinite, of which kaolinite is the most common. It forms 'books' of multilayered crystals and results chiefly from alteration of feldspars by hot fluids or gases or by weathering. The white conical tips at St Austell in Cornwall, England, are waste from extracting kaolinite weathered from granite. The white clays of the south-eastern United States are also kaolinite. When a deposit of kaolinite is pure and free from iron coloration, having been reworked, it is often termed **pipe clay**. China clay, used in pottery, consists mostly of kaolinite.

SMECTITE The common smectite is **montmorillonite** which can also be used more generally as an alternative name for this group. The smectites occur as an alteration product of volcanic ash and tuff. The rock, **bentonite**, common in the western United States, and **Fuller's earth**, found in the Jurassic rocks of the south Cotswolds in England, are both composed mostly of montmorillonite. Smectites have the ability to absorb water and swell, and because of this, smectites are often called swelling clays. Other less common smectites are, nontronite, saponite, and beidellite.

Wadhurst clay

pipe clay

Fuller's earth

London clay

Kimmeridge clay

5cm

OTHER CLAY MINERALS These include chlorite, glauconite, sepiolite, and palygorskite.

Chlorite and glauconite are both green in colour and can often form larger crystals which are visible to the naked eye. Chlorite is a metamorphic mineral but may also occur as a result of the reaction of other clay minerals with iron and magnesium. Glauconite is related to illite and the micas but has a substantial amount of iron; it is the only iron mineral which seems to be forming at the present day on the sea floor, occurring in the form of small pellets.

Concretions

Concretions are spherical or flattened bodies, often occurring in mudrocks; they may be small or, often, up to a metre or so in diameter. Iron concretions may often coalesce to form continuous beds. Ironstone beds of this type occur in the Wadhurst Clay of the Weald in south-east England.

Concretions may be formed as an intermediate stage of lithification. They form when cement grows outwards from a nucleus and, if there is an insufficient cement mineral supply, the rock remains partly unlithified and the concretions can weather out from the softer surrounding material. Some concretions, such as the spherical concretions of the Magnesian Limestone of Durham, England, are so regular and localized that more study on their origin is required.

Septarian nodules are concretions whose processes of formation are still poorly understood. On page 141 there is an example which weathered out from marl and clay deposits at Charmouth, Dorset, England.

Shales

When flat, disc-shaped minerals, such as clays and micas, are compacted, the plates realign resulting in a deposit called shale that flakes along one preferred plane of orientation. Unlike claystones shales do not form a plastic mass when wet although they may disintegrate when waterlogged. Shales are not necessarily directly derived from clays and are often very different in composition. Shale often has a higher content of larger silt-sized particles. Sedimentary variation in particle type can lead to laminated sediments with fine alternations of clay-rich and silt-rich material. This variation can lead to differing degrees of flakiness.

Tilestone is a type of shale where the rock breaks in tile thicknesses; when the amount of mica decreases, a shale may form thicker bedding and become a **flagstone** which is usually more properly considered a silt or sandstone. Very thinly bedded shales are occasionally termed **paper shales**. When a shale is subjected to increases in pressure, by burial for example, it can become a slate.

Siltstone

Siltstone is composed of particles between 0.004 mm (0.00015 in) and 0.0625 mm (0.0025 in). The mineralogy of silt is varied and commonly

marl
bituminous shale
oil shale
interlaminated
silt and shale
shale
fossiliferous shale
5cm

includes quartz, feldspar, mica, and calcite, with little clay. The finer silt, derived originally from desert areas or from vegetation-free areas around ice sheets, can be blown great distances and result in a deposit known as **loess**. When loess has been reworked by river action, it is called **brick earth**.

Other mudrocks

MUDSTONE Like shale, mudstone is not plastic when wet and its cohesion and low water content are similar, but it lacks the flakiness of shale.

MARL Marl is a less common mudstone with a calcareous cement which can be identified by its slight effervescence when white vinegar is applied.

ROCK FLOUR This is rock material that has been ground to a uniform clay or silt during faulting, when it is termed **fault gouge**, or by glacial action where rocks embedded in ice are ground against each other and the bed rock. When glacial rock flour is transported by meltwater, deposits known as **till** and **glacial drift** are deposited. These can also contain whole rock fragments and boulders of great size.

Impurities

Even a very low percentage of impurities, such as iron, can impart a strong coloration to a marl or claystone. The form of iron oxide which may occur in waterlogged soils will impart a blue-green or grey colour to the rock while that which occurs in desert conditions will stain the rock red.

The presence of certain impurities in clay enables it to be used for specific purposes. **Brick clay**, for example, contains substances that promote fusion to produce good bricks; **fireclay** (once a soil and often termed **seat earth**) has a very low content of such substances and is used for making refractory bricks for lining furnaces and chimneys.

Organic matter in clays and shales can give rise to **bituminous** or **oil shales**. An organic-rich rock will often smell oily when freshly fractured. In low-oxygen environments, such as in stagnant hollows on the sea floor like those in Norwegian fjords or the Black Sea, evil-smelling black muds may be deposited. The black shales of the Lower Palaeozoic in Britain are an example of such a deposit, coloured black by iron sulphides, such as pyrite, and carbon.

Fossils

The fine-grained nature of this sediment often results in good fossil preservation. Often the fossils are preserved by pyrite which tarnishes when exposed to the air. Pyritized ammonites and belemnites may be found as concretions in the Black Venn marls and shales at Charmouth on the south coast of England. Plant remains are also often preserved, particularly in anoxic (lacking oxygen) and reducing (removes oxygen) sediments because few bottom dwelling fauna could inhabit such an inhospitable environment.

glauconitic marl

graptolitic shale

pyritic shale

pyritized ammonite

shale with
trilobite fossils
5cm

Sandstones

Essentially, sandstones are classified on the basis of the average relative proportions of their three main components, quartz, feldspar, and rock fragments. It is possible to make the assessment using a hand lens.

QUARTZ This is the most common mineral in sandstone and the one most resistant to weathering and transport. As we have already discussed, quartz grains can readily be identified using a hand lens (*see* page 60). Most quartz grains are derived from granitic rocks, acid gneisses, and schists but very clear (inclusion-free) crystals may indicate a volcanic origin. Quartz, derived from veins and fractures will be milky white due to **fluid inclusions**. Often, previously formed sandstone can weather and yield quartz fragments that have already undergone at least one **sedimentary cycle** (*see* page 26).

FELDSPAR Feldspars are softer than quartz, less chemically stable, and have two good cleavages; they are also less resistant to weathering (*see* page 60). They are sometimes pink or greenish grey in colour, and usually more rounded than quartz grains in the same rocks. They often show shiny cleavage surfaces. As feldspars weather and alter they develop an increasingly dusty appearance, eventually altering and breaking down into sericite (a variety of muscovite), kaolinite, and illite. Feldspars are derived from the same rocks as quartz and usually only last for one sedimentary cycle before they break down into clays.

ROCK FRAGMENTS Rock fragments predominate in coarser sandstones, **conglomerates** and **breccias**. As the rock fragments become smaller, they tend to break down into their constituent minerals. Rock fragments derived from within the area of deposition may often consist of lumps of mud that bend around other components of the sandstone when compacted. Fragments derived from outside the area of deposition may have been transported great distances. Other, softer, clasts will tend to break down more rapidly.

Colour

The colour of a sandstone can be a misleading feature because it is controlled by small traces of chemicals, such as iron oxide. The weathered surface of a rock is often a totally different colour to that of a freshly broken one. The weathered colour often reveals an iron content not otherwise evident from the fresh surface of the rock. The colour of a rock depends upon the chemistry of the cement and upon that of its constituent minerals. A sediment or mineral rich in aluminium, sodium, potassium, calcium, magnesium, or barium will generally be white or fairly light in colour. Sediments and minerals rich in chromium, manganese, cobalt, nickel, titanium, and vanadium will often be deeply coloured. Red Penrith Sandstone, for example, is coloured by iron hydroxide, while yellow Hopeman Sandstone is only lightly coloured and is a fairly pure **quartz arenite** with few trace elements colouring the cement.

millstone grit

arkose

greywacke

fossiliferous
pebbly sandstone

5cm

quartz arenite

lithic arenite

Quartz sandstone

If a sandstone has less than 15 per cent matrix, it is called an **arenite** and such a rock with over 95 per cent quartz is called a quartz arenite. These are the most mature (*see* page 30) of all the sandstones. The grains are mostly well rounded, well sorted and, in many cases, have undergone much reworking and experienced multiple sedimentary cycles. Quartz arenites are often deposited on shallow marine shelves and are well developed in the Cambrian and Ordovician rocks of north-west Europe and eastern USA. Quartz arenites are common throughout the geological record. The Permian Penrith Sandstone, for example, from Cumberland, England is a quartz arenite which often has a red coloration. When a quartz arenite is particularly well cemented, especially with silica, the rock becomes extremely resistant, like sarsen stone, and forms distinctive ridges such as the Stiperstones (Ordovician) of Shropshire in the Welsh/English borderlands. Such a rock is termed a sedimentary **quartzite** (as opposed to metamorphic quartzite). When the amount of matrix exceeds 10 per cent, the rock is termed a **wacke**. **Quartzwackes** have mostly stable constituents and are more compositionally mature than other wackes.

Arkose

Rocks containing more than 25 per cent feldspar, with quartz, and less than about 50 per cent lithic fragments are termed arkoses. Where the feldspar is less than 25 per cent but greater than 10 per cent, the rock is called a **feldspathic arenite** or **wacke**. Arid desert conditions diminish chemical breakdown of the feldspars so that the feldspar clasts are generally fresh, though there may be a degree of alteration to kaolinite and sericite. Arkoses are often red or pink, due to the colour of the feldspars and, often, because of haematite cement which is common in desert environments. Rapid deposition also prevents intensive weathering of feldspars so river environments favour arkose formation.

Lithic sandstone

This is a sandstone where the rock fragment content is greater than that of the feldspar content. In a **lithic arenite** there is very little matrix and the overall mineralogical and chemical compositions are widely varied, dependent upon the abundance and type of rock fragments. This immature composition is due to high rates of sedimentation and relatively short transport distance. **Lithic arenites** are common in river and deltaic environments. Lithic arenites are often called **subgreywackes** because of their roughly similar composition to greywackes.

GREYWACKE This rock type is a lithic sandstone which consists of fine to coarse, angular to subangular particles, most of which are rock fragments. These are dark-coloured, hard rocks that have been deeply buried and were, characteristically, deposited in a rapidly subsiding basin. The matrix is generally composed of a hardened mixture of micas and clays, approximately the composition of slate. Deep burial, with a degree of recrystallization, renders greywacke fairly resistant. The origin

quartz arenite

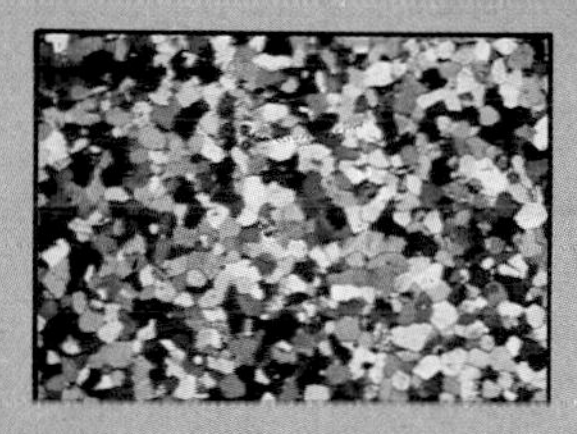

sarsen stone
(x12)

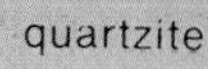

quartzite

5cm

of the matrix is uncertain because modern deep-sea sediments, from which greywackes form, do not contain much mud. Greywackes occur in beds ranging from a few centimetres to a metre or so in thickness. Greywackes are common in the Lower Palaeozoic rocks of Wales, Ireland, and the Southern Uplands of Scotland. Well-bedded greywackes outcrop in Cardiganshire, Wales.

Other sandstones

Many sandstones are characterized by a relative abundance of a specific mineral other than quartz or feldspar. Some of these are fairly common and have specific sedimentological associations. (*See* page 129.)

GLAUCONITIC SANDSTONE Often called greensand and contains enough of the green mica glauconite to impart a green colour to the fresh unweathered rock (it is brown when weathered).

MICACEOUS SANDSTONE When a sandstone has a reasonable content of mica, these platy minerals can cause the rock to split readily into flagstones from about a millimetre to several centimetres thick, depending upon the distribution and concentration of the micas. The surface of such a flagstone will sparkle and, under a hand lens, the individual mica flakes will be clearly visible.

PHOSPHATIC SANDSTONE Low sedimentation rates can lead to the accumulation of otherwise rare particles, such as phosphatic bone, teeth, scales and coprolites (faecal pellets) of fishes and other animals, to form bone beds. These deposits are often mixed with a variety of clastic and carbonate minerals and are not particularly common. Tertiary bone beds exist in North Africa and the Middle East, where they are an economic source of phosphate for use as fertilizer. The Ludlow Bone Bed is a well-known phosphate-bearing horizon found in the Welsh borderlands. Most outcrops have been ruined by over-zealous collection by geologists but some samples can be obtained by carefully searching stream beds and tracks that cut into this horizon. Take care not to hammer actual outcrops.

Cements

ARGILLACEOUS CEMENTS An argillaceous or clay-based cement may occur either as a result of an initially very poorly sorted rock or due to the alteration of chemically unstable constituents. A rock with such cement is not very resistant unless a degree of recrystallization has occurred. The Sandgate Beds in West Sussex are an horizon of **argillaceous sandstones** within the Lower Greensand which have been selectively eroded by the River Rother due to their soft argillaceous matrix.

CARBONATE CEMENT This, together with silica, is the most common type of sandstone cement. Calcite is the most common carbonate to occur as a cement in sandstone. Dolomite and other carbonates, such as siderite, are also fairly common. There are two main types of calcite cement, drusy spar and poikilotopic crystals. A sandstone cemented by calcite (**calcareous sandstone** – *see* page 129) is more easily weathered owing to its relative softness and solubility in acid.

arkose

arkose
(x12)

lithic sandstone

lithic sandstone
(x12)

greywacke
showing graded
bedding

greywacke
(x12)

glauconitic
sandstone

5cm

SILICA CEMENT Cementation by silica occurs in two main ways. Deep burial can force the quartz grains together, welding the grains at points of contact and dissolving quartz which is then redeposited in the pore spaces. Alternatively, the silica may be precipitated by silica-rich fluids that percolate through the rock. Silica cement is precipitated around quartz grains, and can make the rock extremely resistant although impurities or incomplete cementation can weaken them. Penrith Sandstone is a quartz arenite in which the grains are coated with iron hydroxide, weakening the bonding of the grains to the cement matrix.

IRON CEMENTS Iron cements often occur in rocks deposited in desert environments. They can form a coating around the grains, loosely bonding the rock together; the addition of silica will bond the grains more tightly. Iron pans are iron-cemented layers in loosely consolidated sediment. These iron pans can result in lenses of cemented rock (**ironstone**) that form outcrops in poorly cemented sandstones.

Breccia

Breccias need not necessarily be composed of siliceous material. Most breccias are closely related to the rocks from which they were derived. Scree slopes in Carboniferous Limestone areas are good examples of limestone breccias caused by weathering. The fault breccia found in the Brora region of north-eastern Scotland consists of large angular fragments up to 45 m (150 ft) in size. This monstrous deposit may well have formed by earthquake action on an undersea cliff.

Other breccias (intraformational) can be caused by the solution of evaporites resulting in the collapse of overlying rocks, as seen in the Permian dolomite on the Northumberland coast near South Shields, England.

Rocks are shattered during earthquakes which occur when built-up tension in rocks is released. Often, movement along faults occurs and shattered rock builds up in the space between the two surfaces. Such a rock is called a **fault breccia** (*see* page 131).

Conglomerates

Conglomerates are coarse, clastic rocks composed of subrounded to rounded particles. These rocks are often subdivided on the basis of the number of different pebble types and also upon the origin of the component clasts. Textural variations provide other classifications, such as grain or matrix support, and the size range of individual clasts. Often more specific names are given according to individual characteristics. **Puddingstone** is named for its textural similarity to a fruit pudding; it is composed of well-rounded pebbles in a white sandy matrix. A conglomerate need not necessarily consist of sandstone fragments, and may be composed of chalk clasts, for example.

TILL AND TILLITE Till is a deposit formed more-or-less directly as a result of the grinding action of glaciers. Rock surfaces are ground to flour by rocks imbedded in the base of a glacier. This rock flour is often carried away by streams that flow under the glacier. When the glacier

glauconitic sandstone with chalk clasts

phosphatic sandstone

poikilotopic crystals (x12)

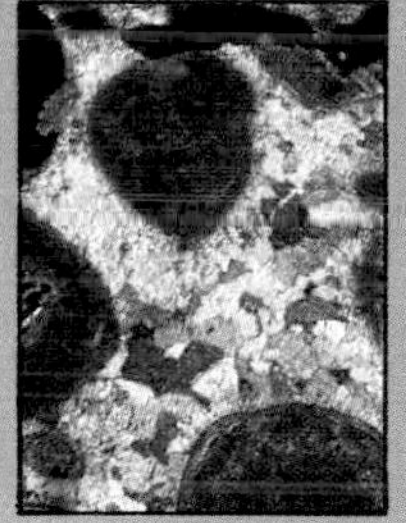
drusy spar (x12)

calcareous sandstone

micaceous sandstone

argillaceous sandstone

5cm

ironstone

iron-cemented sandstone

melts, the rock fragments carried in the glacier are dumped, resulting in a deposit of a half to two-thirds rock flour and the remainder consisting of rock fragments from gravel to boulder size. Tills are well developed in East Anglia, England where thicknesses over 60 m (200 ft) are known. Tills are also common in North America, for instance; there is a good Pleistocene till at Bull Lake, Wind River, Wyoming. Tillites are hardened tills, often many millions of years old; for example, the Sturtian Tillite near Adelaide, Australia.

Heavy minerals

If you look carefully at loose sand, especially river sand, you should be able to see a very few (less than 1 per cent) coloured grains among the quartz and feldspar. These are accessory minerals and can be very resistant to chemical weathering and mechanical abrasion. These are called heavy minerals because they have a higher specific gravity than quartz and feldspar. The common non-opaque minerals are zircon, tourmaline, rutile, apatite, garnet, staurolite, and epidote. Common opaque minerals are pyrite, magnetite, and ilmenite.

ILMENITE This is an oxide of iron and titanium. It is usually black with a metallic lustre, often appearing as elongated grains, irregularly shaped, and sometimes in masses. Ilmenite may be distinguished from magnetite or haematite by its tendency to form skeletal crystals and by the white alteration product, **leucoxene**.

HEAVY MINERAL SEPARATION Heavy minerals are commonly separated by using a heavy liquid, but it is more convenient to do it by **panning** in the way that prospectors extract gold from rivers. Panning can be done by using a frying pan and if the handle can be removed, so much the better. The technique takes some practice but you may even find gold! Put about a tablespoonful of sand or crushed rock into the pan with some water; tilt the pan slightly, until the water just overflows. Swirl the pan gently in one direction, keeping the sand in the angle of the pan on the side furthest from you. As you swirl the darker minerals will gradually form a 'tail' in the pan. Scoop this tail out, dry it on some blotting or filter paper, and examine it. The heavy mineral grains can be difficult to identify owing to abrasion and alteration.

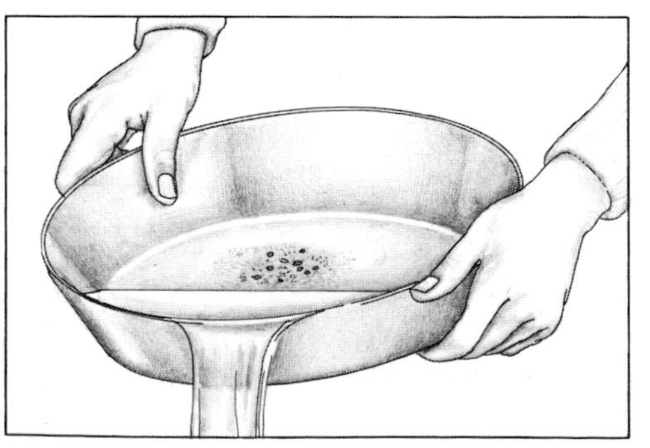

Separating out heavy minerals using a shallow pan.

conglomerate

5cm

Limestones

Generally, limestones are any rocks in which the proportion of carbonate material, such as calcite, exceeds the rest. They effervesce vigorously with dilute hydrochloric or white vinegar. The two most important constituents are calcite and dolomite but small amounts of iron-bearing carbonates, such as siderite, may also occur. In modern limestone sediments, the two different crystal forms of calcium carbonate, calcite and aragonite occur.

CALCITE This is generally colourless with a vitreous lustre but is often cloudy or coloured by impurities. This mineral often occurs as very fine crystals, mostly less than 0.004 mm (0.00015 in), usually termed **micrite** which is an abbreviation for **micro-crystalline calcite**. Crystals coarser than micrite are termed **spar** or **sparite**. These may form as a cement and also grow in veins and cavities as **dog-tooth** or **nailhead calcite**. Calcite has a hardness of 3, a white streak, and perfect cleavage in three directions intersecting at 75°. It may be distinguished from dolomite by staining, its distinctive crystal forms, and by its far more vigorous reaction with dilute hydrochloric acid or vinegar.

ARAGONITE This mineral is relatively rare in sedimentary rocks because it alters easily to calcite, the most stable form of calcium carbonate. It occurs mostly as a secondary mineral in cavities of basalts and andesites. Aragonite usually shows a columnar, fibrous structure with a six-sided cross section. Cleavage is imperfect and parallel to the length of the crystals. Aragonite does not cleave as easily as calcite and is very slightly heavier and harder (hardness 3½). It was probably once a widespread constituent of limestones but has since altered to calcite.

DOLOMITE This is a colourless mineral forming fine- to coarse-grained aggregates. Crystals are often well-shaped rhombs. Dolomite usually occurs as a diagenetic alteration (*see* page 138) of limestone. Often, the original texture of the rock will be totally obliterated by this alteration although 'ghosts' of the earlier fabric may still be recognizable.

SIDERITE This is a rare and generally minor constituent of some limestones. Iron is commonly present in the mineral dolomite but, in a few cases, it occurs as scattered sideritic rhombs. Slight oxidation results in the breakdown of siderite which is then readily detected by heavy iron oxide stains along the cleavage and grain boundaries.

Classification of limestones

Three schemes are currently used for limestone classification, using grain size, grain relationships, or dominant components. Here, we have used the scheme that combines dominant components with matrix type, producing names such as **oosparite** and **oomicrite** (*see* page 134). Components of limestone may be divided into four groups: micrite; sparite and other cements; non-skeletal grains; skeletal components or fossils.

5cm
biomicrite
biosparite

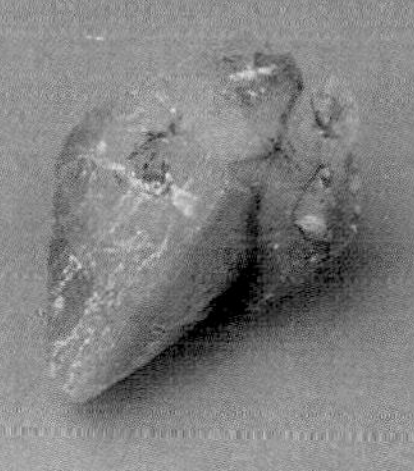
dog tooth calcite

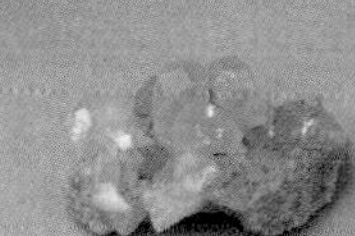
nailhead calcite

aragonite

coarse dolomite
finely
crystalline
dolomite

Micrite

This is microcrystalline calcite, usually comprising calcite crystals generally less than 0.004 mm (0.00015 in) across. Often, diagenetic alteration can coarsen these crystals to **microspar** which is a term used to describe calcite crystals from 0.005 to 0.010 mm (0.0002 to 0.0004 in). A micrite that is uniformly fine and compact is often called **lithographic**, because such rocks were once used as lithographic printing plates. The best-known example of such a rock is from **Solenhofen** in Bavaria. Micrite probably originates from the disintegration of calcareous green algae although it may possibly precipitate directly from sea water. A **biomicrite** is a micrite containing fossils.

Sparite and other cements

When there is little or no micrite present in a limestone, the space between the grains is often filled by calcite cement. This form of calcite is known as spar or sparite. The crystals can form different textures together with other cements.

Non-skeletal grains

OOIDS These are often referred to as ooliths and are spherical to subspherical grains with one or more regular layers arranged concentrically around a nucleus, often a fossil fragment or a quartz grain. These concentric layers are often visible to the naked eye on broken surfaces. Ooids are typically 0.2 to 0.5 mm (0.008 to 0.02 in) across. Today, ooids form in agitated waters where they occur as dunes and sand waves, usually in water depths of about 5 m (16 ft) but they may reach between 10 to 15 m (30 to 50 ft). When a nucleus or larger fragment has only one layer it is called a **superficial ooid**. A rock composed almost entirely of ooids is called an **oolite**, or, more specifically, an **oosparite** or **oomicrite**.

Sometimes ooids may form clusters which themselves act as nuclei and form composite ooids. Under certain circumstances, such as in hot springs and in some cave streams, ooids may sometimes measure 1 cm (0.5 in) across; these are known as **pisolites** or **cave pearls**. Oolitic limestones can be found all over the world; the oolite from Pennsylvania, USA, has no cement. The Inferior and Great Oolites of the Jurassic, outcrop in a band from the Dorset coast in the south of England to Scarborough in the north-east.

The origin of ooids is not known for certain. Some geologists believe them to be formed by direct precipitation from sea water, while others believe that their origin is involved in some way with bacteria or, possibly, algae.

PELOIDS These are spherical, elongate or angular grains composed of micrite. Many peloids are faecal pellets or coprolites of sea creatures; these are usually ellipsoidal. Others are often of indeterminate origin. They may be **intraclasts** (*see* page 136) or probably **micritized** fossil fragments. Fragments lying on the sea floor may be bored by **algae**. When the algae dies, it leaves a microscopic tube that becomes filled by

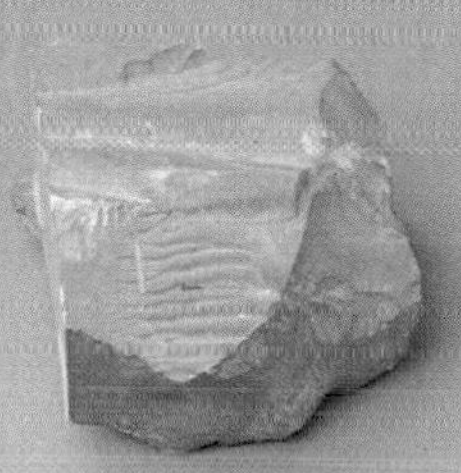

micrite
(Solenhofen limestone)

biomicrite

biosparite

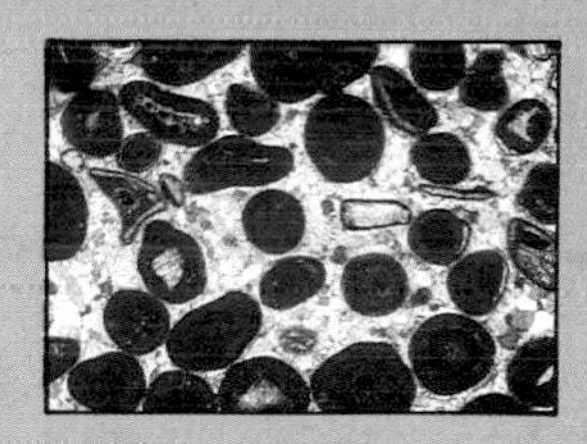

oolite
(x12)

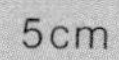

siliceous oolite

pisolite

micrite which hardens. Algal bores criss-cross each other and fill with micrite until the bored fragment is buried by other sediment. When this process is only partial, a **micrite envelope** is formed; when the fragment is totally bored and micritized it is a peloid.

INTRACLASTS These are fragments of lithified or partly lithified sediment. A common type of intraclast in limestones is a micrite flake derived from desiccated (dried-out) tidal flats. An abundance of these flakes produces a **flakestone** or **edgewise conglomerate**.

Skeletal components or fossils–bioclasts

A wide variety of fossils, both whole and fragmented, occur in limestones. The major groups of fossils that contribute to limestones are described below.

MOLLUSCS AND BRACHIOPODS These are sea creatures with exterior shells. Molluscs include **bivalves, gastropods**, and **cephalopods**. Bivalves are a large group of species with two shells or valves usually of equal size and tolerate a wide range of environments. Certain bivalves, such as oysters, form reef-like banks while they live. During the Cretaceous Period a coral-like bivalve, called a rudistid, formed reefs in Mexico, southern Europe, north Africa, the Middle East, and elsewhere. Often, empty mollusc shells are reworked by waves and form banks. On the opposite page is an example of shelly limestone (biomicrite) from the Isle of Wight, England. Gastropods can also form reef-like structures but mostly these bottom-dwelling creatures browse in soft sediment or prey upon other creatures, such as bivalves. Gastropods occur in vast numbers in restricted salt lagoons and in brackish water because they are able to tolerate fluctuations and extremes of salinity. Cephalopods, such as ammonites tend to be far less abundant in sediment because they are free swimming and consequently occur over a far wider range of environments.

Brachiopods are less common now than they were before the Tertiary era when they occupied similar environments to bivalves.

ECHINODERMS These are marine organisms which include echinoids (sea-urchins) and crinoids (sea-lilies). Echinoderms, particularly the crinoids, are major constituents of many limestones of the Palaeozoic and Mesozoic. The Carboniferous Limestone of northern England is often rich in crinoid debris. This rock is familiar as a decorative facing stone. Echinoderm fragments are easy to recognize because they are composed of large single calcite crystals. In hand specimen single crystal fragments usually show a smooth cleavage face. A syntaxial cement often forms around these fragments.

FORAMINIFERA These are predominantly microscopic and not visible in hand specimen. The floating varieties dominate some open marine deposits, such as in the **Globigerina ooze** of ocean floors, and some Cretaceous and Tertiary chalks. Bottom-dwelling foraminifera are common in warm, shallow seas. A relatively large, up to about 5 cm (2 in), bottom-dwelling foraminifer is *Nummulites*. This coin-shaped creature formed large banks in water depths of probably 10 m (30 ft) or

intrasparite
shelly biomicrite
gastropod
brachiopod
bivalve
syntaxial cement
(x12)
echinoid in chalk
5cm
crinoid stems
in limestone
(crinoidal limestone)

so. The distinctive **nummulitic limestone** occurs throughout the Mediterranean region and was used as a building stone by the Romans, and several of the Egyptian pyramids incorporate this rock. Nummulites are also found in the Eocene Bracklesham beds in southern England.

ALGAE Algae make an important contribution to limestones by providing skeletal carbonate grains, by trapping grains to form **stromatolites**, and attacking and boring rocks and particles yielding micrite grains. Four groups of algae are important:

1 Red algae can form calcareous skeletons which often encrust surfaces such as the sea bed and reefs. When shells or pebbles are encrusted, **rhodoliths** develop. These algae are important in binding and cementing loose fragments and sediment and, in some cases, participate in reef formation.
2 Green algae are important contributors of micrite particles.
3 Blue-green algae are important for stromatolite formation, and in micrite envelope and peloid formation. Some blue-green algae form nodules (**oncolites**) with concentrically arranged, often asymmetrical layers. The **oncolitic limestone** illustrated is from Morocco.
4 Green-yellow algae are so small they can only be studied effectively using a scanning electron microscope. These algae are called **coccoliths** and form a significant component of modern deep-water carbonate oozes and is common in Tertiary and Cretaceous deposits. The **chalk** extending across the south, east, and north-east of England and much of north-western Europe is composed predominantly of coccoliths.

Other fossil fragments include: bryozoans, which are small, colonial marine organisms; corals which can build reefs and also contribute detrital fragments; arthropods, such as trilobites and sponges contribute calcitic and siliceous spines and also bore into substrates providing micrite grains.

Diagenesis

The onset of diagenesis is often difficult to define in limestones. Non-diagenetic processes, such as algal micritization, may be occurring at the same time as diagenetic processes, such as cementation or **dolomitization**.

DOLOMITES These rocks are composed primarily of the mineral dolomite. On the basis of their dolomite content carbonate rocks may be divided into: **dolomitic limestone** (10 to 50 per cent dolomite); **calcareous dolomite** (50 to 90 per cent dolomite); and dolomite (90 to 100 per cent dolomite). The origin of dolomite is still a major problem in geology. Basically, the element magnesium, common in sea water, is introduced into the molecular structure of calcite, possibly during mixing of fresh water and sea water. The majority of dolomites are replacements of pre-existing carbonate sediments where fluids, rich in magnesium, flow through the pores and change the chemical composition of the rock. Some very fine dolomites may well be primarily precipitated as evaporites. Dolomite forming almost at the same time as

pelagic mud
(x12)
oncolitic limestone
nummulitic
dolomite
nummulitic
limestone
crystalline dolomite
partially dolomitized
limestone
5cm
'cannonball limestone'
(dedolomite)

primary dolomite is difficult to distinguish but may be recognized by its close association with intertidal and supratidal features, such as fenestrae or birds' eyes which are cavities probably caused by gas trapped in the sediment, stromatolites, desiccation cracks, and flakes caused by drying out of the tidal flats and evaporites. Many of these associations can be observed in modern sediments where dolomite is forming today; in the Arabian Gulf, the Bahamas and Florida.

Dolomitization may also occur late, long after a limestone has become a rock, sometimes when it is exposed at the surface. Illustrated on page 139 are some dolomite specimens showing a range of original fabric preservation. The **nummulitic dolomite**, from Tunisia, is completely dolomitized yet still shows individual fossils. The other two examples show partial and total obliteration of the original texture.

Occasionally, the reverse of dolomitization, dedolomitization, can occur. This process causes strange textures and may be associated with unconformities. The **dedolomite**, also known as the **Cannonball Limestone** is from north-east England (*see* page 139).

CEMENTATION In limestones this involves the precipitation of calcite in pores, and some of the coarser textures may be visible with a hand lens. **Hollow calcite nodules** are illustrated opposite where the calcite spar is coarse and the individual crystals are visible with the naked eye. The banded specimen shows how colour can vary according to the chemical content of the fluids which precipitated the calcite.

Cementation can occur after very little burial or even in exposed sediments in the surf zone on a beach. The limestone formed in this way is called **beach rock**. Beach rock may be found forming today along some parts of the Mediterranean coast. Often, after early cementation, a limestone is buried more deeply and a second cement infills the pores.

Cement types and textures are similar to those found in sandstones. For example, compare the cement in sandstone on page 129 with the cement around crinoids (*see* page 137).

NEOMORPHISM In this process the fabric of a rock recrystallizes without dissolution. Micrite-sized calcite can become coarser and possibly resemble sparite. Fossils can be neomorphosed and lose their original texture by recrystallizing to clear crystals. Neomorphic recrystallization is difficult to distinguish from cementation, although ghosts of the original fabric may still be evident in the neomorphic crystals.

Staining

This process helps to identify the different constituents of limestone. In essence the method is as follows. A very smooth surface is prepared on a limestone sample. This surface is dipped in a solution of Alizarin Red S, potassium ferricyanide and dilute hydrochloric acid, and then rinsed with distilled water. Calcite will be stained red, whilst dolomite, siderite and non-carbonates will be left unstained. Iron-rich calcite will stain blue or purple and iron-rich dolomite will be light to dark turquoise. The stain is not permanent, but may be replicated by taking an acetate 'peel' which can then be examined with a lens; some examples are shown opposite.

banded calcite

stained septarian nodule with calcite veins

nodule with spar

5cm

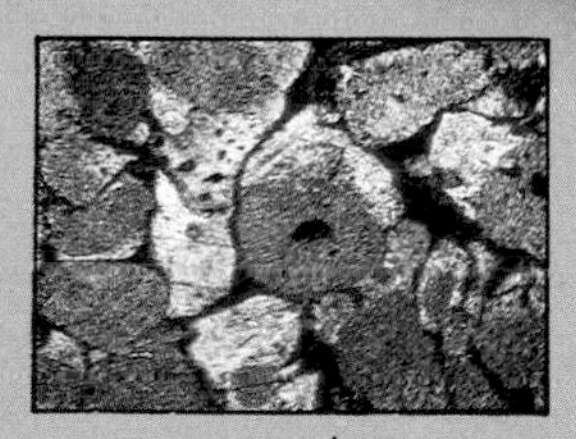

limestone peel (x12)

Evaporites

Evaporites are frequently associated with carbonates and are mainly chemical sediments formed by precipitation from water in which the dissolved salts have been concentrated by evaporation.

Common evaporite minerals

GYPSUM Gypsum mostly occurs as nodular masses within mudrocks and fine dolomites or as closely packed nodules known as **chicken-wire texture**. Layers and beds of gypsum occur which may be contorted. These types of occurrences are typical of sabkha (supratidal) deposits and are often associated with other intertidal and supratidal features (*see* pages 37 and 38). Gypsum may also grow by replacing rather than displacing carbonate sediment forming large crystals. Gypsum often grows in soil overlying gypsum deposits and may form fibrous masses of **satin-spar**, **swallowtail-** or **fishtail**-shaped crystals, **crystal aggregates** and **desert roses**. These samples of gypsum in different forms (*see* also page 145) can be found across north Africa and the Middle East wherever gypsum rocks outcrop at the surface. When gypsum is heated and ground to a fine powder it is known as plaster of Paris, as that was where it was first made.
ANHYDRITE This forms when gypsum is buried to 200 m (650 ft) or so. Water leaves the gypsum molecular structure and forms anhydrite. When uplifted and exposed at the surface, the reverse process occurs. Anhydrite may also be precipitated directly as nodular masses much like gypsum, usually in the parts of a sabkha closest to land.

Gypsum and anhydrite can be reworked from the supratidal and intertidal environments and deposited as a 'clastic' sediment. Horizons of small- and large-scale folded, contorted, and brecciated gypsum and anhydrite also occur and are generally interpreted as slumps, slides, and debris flows. Examples of reworked gypsum and anhydrite occur in the Elk Point Basin, Canada, the Permian deposits of north-east England, Denmark, and northern Germany.
HALITE This is also known as **rock salt** and is a massive, coarsely crystalline mineral without joints; in some deposits it is layered. Salt may occur as alternate layers with dolomite or other evaporites; it may also show a layered variation in colour or translucency caused by various inclusions in the crystals. The salt crystals may be well-defined cubes or, possibly, hopper shaped. Salt is a rock which is prone to flow at fairly low temperatures and pressures. It may rise from a deeply buried stratum, to pierce overlying strata and cause domes. These features are common on the Gulf Coast of Texas and are also known in Germany, Russia, the Middle East, and north Africa where they emerge at the surface to give circular areas of white, salty soil often surrounded by a ridge of hills. Many other mountains and hills in these areas may be the result of rising salt. It can also rise as a 'wall' which may result in a long anticline. Often these salt domes and other evaporites form oil traps by forming an

5cm

anhydrite

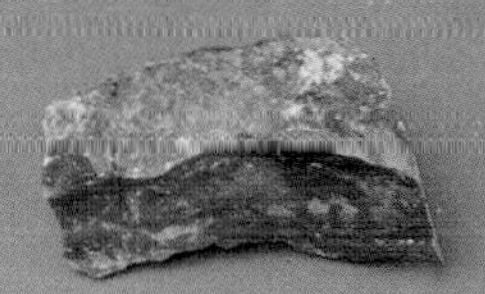

laminated dolomite

anhydrite with
darker dolomite

fine gypsum rock with
large gypsum crystals

fibrous gypsum

gypsum

gypsum replaced
by calcite

impermeable cap and preventing the oil from migrating upwards (*see* page 185 for illustration).

Other evaporites

Other potassium and magnesium salts, such as sylvite, carnellite, and glauberite are found, but these are much more soluble than gypsum or anhydrite so that they are less common. They are of great economic importance, however. The very soluble salts are called **bitterns**.

Dolomite is also frequently noted in evaporite deposits (*see* pages 132 and 138).

BARITE This is found in nodular masses in soils occasionally forming desert roses (for example in the sands of Oklahoma) similar to gypsum but can easily be distinguished by its greater specific gravity. Barite is usually colourless to white but can be grey, yellow, or brown. It also occurs as **crystalline vein fillings** as shown in the specimen.

STRONTIANITE This occurs in similar situations to barite but is generally less common. It is a carbonate of strontium with a white streak and is generally grey, yellowish, or greenish in colour. An example of **strontianite with calcite** is illustrated. These beds may be the result of the dissolution of evaporite minerals and the subsequent collapse of overlying rocks.

Occurrence

The Permo-Trias basins in Cheshire and north-east England, and in Holland and northern Germany, contain great thicknesses of evaporites, usually gypsum, anhydrite, halite, and occasionally the more soluble potassium salts. The total thicknesses of these evaporites are 400 to 500 m (1300 to 1600 ft). Such enormous deposits must have formed where an immense gulf was undergoing continuous evaporation with periodic replenishments from the open sea. Theoretically, evaporation in such a basin should result in a sequence with the least soluble gypsum and anhydrite at the base passing up through halite into a range of bitterns. In the Stassfurt Basin in Germany, the deposits show two such sequences although there are some small-scale variations. The isolated Gulf of Kara Boghaz on the eastern side of the Caspian Sea gives a reasonable model of processes of evaporite formation.

It is thought that the Mediterranean Sea was once such a basin and that during the late Miocene it dried up completely. Great thicknesses of evaporites apparently underly the sea bed. The basin would have been periodically replenished by the Atlantic flowing through the Straits of Gibraltar; these straits must have been a vast waterfall at that time.

Evaporites can also form in lakes. The Dead Sea in Israel and the Great Salt Lake in Utah, USA, have predominantly halite deposits. Bitter lakes such as Mono Lake (California) and Carson Lake (Nevada) have predominantly sodium carbonate and sulphate minerals. Ancient lake evaporites are well developed in the Eocene Green River formation of Wyoming and Utah in the United States.

strontianite and calcite

gypsum 'desert rose'

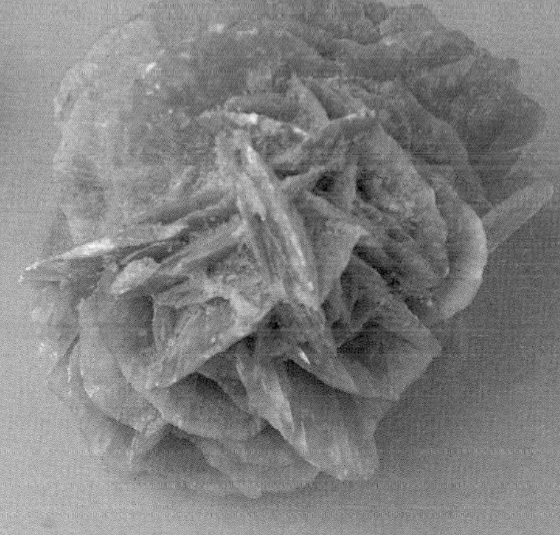

5cm

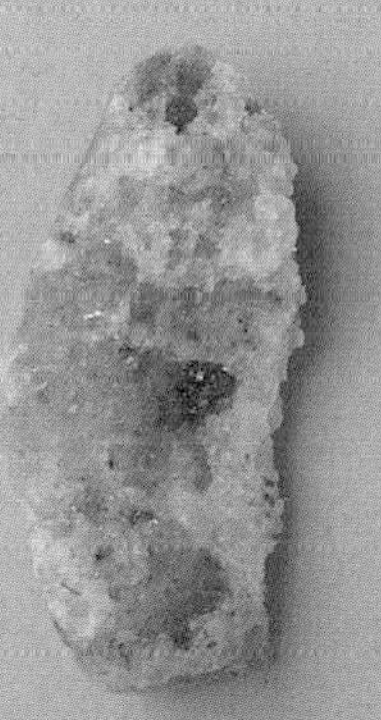

rock salt (halite)

barite

Rocks of organic origin

Strictly, many limestones are of organic origin comprised mostly of shells and other skeletal fragments. Certain specific organisms are responsible for creating quite common rock types.

Reefs

A reef generally refers to masses of organic skeletal material which consist partly of organisms that have grown in their life position, partly of chemically precipitated material, and partly of organic debris which may be derived from erosion of the reef itself or from material that has been transported to the reef.

CORALS These are the most common modern reef-building organisms. Species of coral vary widely in their form and structure. The **tabulate corals**, for instance, are simple colonial corals, now extinct, in which the individual coral wall is the most important structure. Tabulate corals may be found, for example, in the middle Carboniferous Paradox formation in Utah, USA. *Favosites* and *Halysites* are examples of tabulate corals. **Rugose corals** are often solitary, trumpet-shaped corals. These may also be colonial but have never been major reef builders. Scleractinian corals, on the other hand, are typical reef builders of the warmer climates. Reefs occur in the Upper Jurassic Corallian in Cambridgeshire, where *Isastrea* occurs. Corals are not important in the Cretaceous rocks of England, and only a few small *Goniopora* are found in the Eocene Bracklesham beds of the Hampshire basin.

CORALLINE ALGAE This is a popular name for red algae which have calcareous skeletons and are able to build reefs.

BRYOZOANS This group of animals forms colonies but only calcareous types are found as fossils. They are able to build reefs in much the same way as calcareous algae.

OTHER REEF BUILDERS Sponges, serpulid worms, and some gastropods are of limited importance in reef building. Oysters form large structures which may be thought of as reefs. Rudistid bivalves also formed 'reefs' during the Cretaceous. Rudistids may be found particularly in rocks around the Mediterranean region and the Middle East as well as in the USA (for example, the Edwards formation of south Texas).

Organisms and plants may also trap micrite, forming mud mounds. Bryozoans and crinoids (echinoderms with stems) occasionally formed 'meadows' in which mud became trapped between the organisms. Plants could also perform a similar function. Often these mud mounds form the core upon which a framework reef may be constructed.

STROMATOLITES These are mats of blue-green algae upon which micrite and detrital fragments adhere. Continued algal growth combined with the adhering detritus results in the build-up of significant thicknesses of deposit. These occur only along coastlines and are often associated with evaporite formation. Stromatolites occur throughout the geological record but are particularly important in the Precambrian

Favosites
(coral)
Halysites
(coral)
Zaphrentites
(coral)
5cm
Isastrea
(coral)
limestone almost
entirely composed
of oysters
Syringopora
(coral)
rugose
coral
bryozoan
2cm

where thick limestone and dolomite stromatolite sequences occur. Algal mats result in a range of structures from planar and slightly undulating laminations through domed heads, such as occur today at Shark Bay in western Australia, to detached spherical or irregular algal balls known as oncoliths. This range of structures reflects an increasing degree of wave energy to which the stromatolites are subjected during their formation.

Organic matter

Much of the organic matter which comprises living organisms breaks down in the presence of oxygen. When there is a deficiency in oxygen decomposition is incomplete and arrested allowing organic matter to be preserved in the rock. Organic deposits can be divided into two broad groups, those made of material which has been transported or deposited from suspension and those, such as peat, formed by organic growth.

OIL SHALE This is a fine-grained shale containing a largely solid substance which yields oil on distillation. The solids are formed by decomposition of organic material in the shale under oxygen-deficient conditions. The oil shales of the Midland Valley in Scotland have been actively worked in the past to produce oil.

COAL This rock is used as a fuel and occurs as seams in strata of various geological ages. Most coals are formed by accumulation of woody material in place; some, however, are formed from algae, spores, and plant debris which have settled from suspension. In the course of geological time, the heat and pressure resulting from the overlying deposits convert the layers of organic matter into seams of relatively hard coal. Variation in the degree of these processes results in successive stages or ranks of coal ranging from peat, which occurs at the Earth's surface, through soft brown coal (including **lignite**) with visible plant fragments, hard brown coal, **bituminous hard coals** which are black, hard, and bright, and brittle, shiny **anthracite** with a conchoidal fracture.

Lignites and other soft brown coals occur in thick beds on the northern foot of the Hercynian blocks in Germany and between the Rocky Mountains and the Mississippi lowlands in the United States. In Britain they are rare but do occur in some areas, such as the Bovey Tracey basin in Devon. Nearly all the workable coal deposits in Britain are found in Carboniferous deposits. These are generally bituminous coals and anthracite, and occur throughout western Europe and North America.

Cannel has a high volatile content, and is rich in spores, algae and fungal material. **Boghead** consists of algal material together with some fungal matter. **Torbanite** is a well-known boghead coal found in the oil shales of the Midland Valley in Scotland.

PETROLEUM This consists of crude oil and gas. Crude oil is largely made up of hydrocarbons with some oxygen, nitrogen, and sulphur. Too much sulphur makes refining expensive.

Petroleum is derived mainly from the decomposition, heating, and burial of organic matter (often marine micro-organisms). Petroleum can also be generated from coals, but the crude oil produced is more waxy than that derived from marine organisms.

stromatolite
5cm

oil shale

bituminous
coal

cannel coal

anthracite

lignite

Metamorphism of acidic igneous rocks

Acidic igneous rocks contain minerals which are stable throughout most variations of temperature and/or pressure, and are only affected at the higher grades of metamorphism. The fine-grained acidic rocks, such as rhyolites, are the most susceptible to metamorphism.

Thermal metamorphism

Acidic igneous rocks remain largely unaffected by thermal metamorphism apart from situations where the rocks are immediately adjacent to a large igneous intrusion. Biotite and hornblende may recrystallize to chlorite, while quartz and feldspar may recrystallize to form a granoblastic texture within light-coloured, speckled, hard hornfelses. These acidic hornfelses are frequently developed where the acidic rocks are cut by basic intrusions, because high temperatures are involved. The acidic rock may even become melted in places. Examples are rather rare.

Regional metamorphism

This also has only a limited effect on these igneous rocks. Moderate pressures and temperatures may produce **granitic gneisses** which have essentially the same minerals that are found in a granite (quartz, orthoclase, plagioclase, and biotite). The rock has a **gneissose fabric**, however, consisting of quite well-defined dark and light bands. The dark bands are often quite thin, down to 1 mm (0.04 in) and consist mainly of biotite. The much thicker light bands, up to 5 cm (2 in) across, consist mainly of quartz and feldspar and have a granoblastic texture.

Increased temperatures and pressures produce **granulites**. These are different from the granulites derived from the metamorphism of basic rocks. Some of these granulites are called charnockites.

CHARNOCKITES These are rather unusual rocks which are still not fully understood. They can be foliated, but more often have a granoblastic texture or textures like those of the unaltered igneous rocks. Granitic charnockites are dark and contain quartz, orthoclase feldspar, plagioclase feldspar (oligoclase and andesine), and some microcline; the feldspars are often intergrown as perthite. The dark mineral is usually hypersthene. A deep-red garnet, almandine, is also frequently present. In hand specimen, the quartz often has a pale grey-blue colour, while the perthite is brownish grey. Such charnockites are described from the Highland areas of Sri Lanka and the Madras area in India.

Charnockites are generally uncommon rocks and were first observed forming the polished headstone of a grave in an Indian cemetery. The inscription was to Job Charnock, the founder of Calcutta; from this, the name 'charnockite' was coined.

MIGMATITES Further increases in pressure and temperature may produce some localized melting within the metamorphosed acidic rock, producing the dark- and light-banded rocks called **migmatites**, which have complex mixtures of igneous and metamorphic textures.

foliated 'acidic' gneiss
'acidic' gneiss
quartzo-feldspathic schist
'acidic' gneiss
5cm
charnockite
'acidic' gneiss

Metamorphism of basic igneous rocks

Basalts are the most abundant basic igneous rocks and it is the metamorphism of these that is usually considered. The minerals comprising these igneous rocks are rather sensitive to changes in temperature and pressure causing basic igneous rocks to be more readily affected by metamorphism than acidic igneous rocks.

Thermal metamorphism

Thermal metamorphism of basic igneous rocks produces dark, compact, relatively fine-grained, structureless, fairly heavy rocks which are called **hornfels(es)** or, more accurately, **basic hornfelses**. These can be difficult to distinguish from similar, dark-coloured basalts.

LOW AND MEDIUM GRADE At low grades of metamorphism, the basic rocks are recrystallized and consist of amphibole (a variety called **actinolite**) and plagioclase (albite), together with lesser amounts of the green minerals chlorite and epidote, which impart a greenish tinge to the hornfels.

Medium-grade thermal metamorphism, that is, closer to the igneous intrusion, produces fine- to medium-grained dark hornfelses which still contain mainly hornblende and plagioclase (ranging through oligoclase, andesine, and labradorite), in addition to the pyroxene diopside, which may be difficult to identify in fine-grained hornfelses. These rocks are often termed **hornblende hornfelses**. Magnetite, apatite, and sphene are common accessories.

The temperature increases associated with thermal metamorphism which cause recrystallization, gradually destroy the original texture of the igneous rock. Fine-grained igneous rocks are more easily recrystallized than coarse-grained igneous rocks. Thus, at this stage of thermal metamorphism, a dolerite or gabbro may still possess their original igneous texture and be only slightly recrystallized, while a basalt may be completely recrystallized and have a metamorphic texture and a new set of minerals.

HIGH GRADE The highest grades of metamorphism are reached adjacent to the intrusion. Again, medium- to dark-coloured hornfelses are produced. Certain mineral changes also occur. The plagioclase feldspar, labradorite, is found while both hypersthene and diopside occur, although they are difficult to distinguish in hand specimen. The name **pyroxene hornfels** is usually given to these rocks. Some olivine may be present. Accessory minerals include apatite, magnetite, sphene, tourmaline, and spinel. Basic hornfelses possess a **granoblastic texture** and it may be possible to see this texture in hand specimens of coarse-grained, high-grade examples.

Occurrence

Metamorphic aureoles are of limited regional extent and, although basic hornfelses are relatively hard, they exert no great influence on the

actinolite

chlorite

epidote

diopside

hornblende

hypersthene

5cm
'basic' hornfels

landscape apart from any they may have exerted prior to metamorphism. No special techniques are required for their extraction and, as they are susceptible to weathering and alteration, the rusty brown surface layers can be fairly soft, but the interior is hard and resistant when fresh. Examples can be obtained from the Comrie area, Perthshire, Scotland; around Odenwald, Germany; the Sierra Nevada area in California; Vancouver Island, British Columbia; the Cuillins on the Isle of Skye, off western Scotland, and the Tilbuster area, New South Wales, Australia.

Regional metamorphism

The rocks produced by the regional metamorphism of basic rocks vary with the interaction of temperature and pressure changes. Basic igneous rocks are converted into **greenschists** at low temperature and low to medium pressure; to **glaucophane schists (blueschists)** at low temperature and high pressure; into **amphibolites** at medium temperature and pressure; to **granulites (pyroxene granulites)** at high temperature and pressure and into **eclogites** at medium to high temperature and extremely high pressure.

If basic rocks are buried, the first metamorphic changes can be brought about by the pressure produced by the overlying rocks. In buried basalt flows, the metamorphic effects are often located near the upper surfaces or around vesicular areas. The changes are rather subtle and can not be recognized in hand specimen. Labradorite crystals are replaced by the variety of plagioclase called albite but, apart from the schiller effect of labradorite, the two varieties appear very similar in hand specimen. A variety of other new minerals may be present, including epidote, the amphibole actinolite, chlorite, calcite, and even a little quartz. The original textures of the igneous rocks are still commonly preserved.

GREENSCHISTS Generally, these are dark green, fine-grained rocks with a schistose fabric or foliation. The green colour is due to the abundance of actinolite, chlorite, and/or epidote. Quite often, greenschists display a banded or layered appearance, with paler bands richer in plagioclase alternating with darker bands with little or no plagioclase. Sometimes the plagioclase crystals may occur as porphyroblasts so that the rocks are porphyroblastic.

If individual crystals can be recognized, either through a hand lens or with the naked eye, actinolite takes the form of slender prisms and needles which are sometimes aligned in a subparallel fashion giving the rock a lineation. A variety of other minerals can be present in greenschists, depending on the original mineralogy of the basic igneous rock and the subtle variations of pressure and temperature involved. Such minerals include magnetite, sphene, apatite, quartz, biotite (which can sometimes become quite abundant), a mica-like mineral resembling biotite called stilpnomelane, possibly a little muscovite, and occasionally hornblende. Remnants of the original igneous textures may remain.

Greenschists are often located within long narrow zones, called **greenstone belts**, found within the central parts of the larger continental masses, along with metamorphosed sedimentary rocks. Here, the

actinolite

chlorite

epidote

glaucophane

omphacite

jadeite

chlorite schist

actinolite schist

chlorite schist

5cm

glaucophane schist
(blueschist)

metamorphosed basic rocks do not always possess a foliation and can be massive and compact.

GLAUCOPHANE SCHISTS Glaucophane schists or blueschists get their common name from the presence of a blue amphibole, glaucophane, although, generally, they are unremarkable, fine-grained, dark, rarely foliated rocks. Colourless or pale yellow mica flakes may be seen on the surface of these rocks. It is worth remembering that blueschists can be derived not only from metamorphosed basic rocks, but also from metamorphosed sedimentary rocks.

In terms of mineral compositions, greenschists and blueschists are not very different. In blueschists derived from basic rocks, however, lawsonite, pumpellyite, actinolite, garnet (a red variety called almandine), rutile (which replaces sphene), and varieties of clinopyroxene called omphacite and jadeite also occur.

AMPHIBOLITES Amphibolites contain abundant amphiboles (usually hornblende) together with plagioclase feldspar (the varieties oligoclase and andesine) and, consequently, may have a spotty appearance. They also contain garnet (almandine), epidote and biotite, with a little quartz. Amphibolites can also be derived from certain sedimentary rocks but they are more often metamorphosed basic rocks.

Amphibolites can be dark-green, massive crystalline rocks with a poor foliation but with the abundant elongate, prismatic hornblende crystals showing a good lineation, which may be recognized with a hand lens. If rather more plagioclase is present, lighter and darker bands may be developed, while frequent biotite may cause a schistosity or foliation to be developed, and the rocks are then called **hornblende schists**. The deep-red garnets, almandine, often form porphyroblasts, the rock then being called a **garnet amphibolite**. Such rocks are easily recognizable in hand specimen; they are usually dark green with deep red patches. Other minerals include sphene, rutile, apatite, occasionally chlorite, and infrequent diopside.

If the amphibolite has been derived by metamorphism of a dolerite or gabbroic rock, a foliation may be so weakly developed that the amphibolite appears very similar to the igneous rock diorite. Rocks such as these are often termed **epidiorites** and are common in south-western Scotland. Certain epidiorites contain garnet, which immediately distinguishes them from the igneous diorites that contain no garnet. Many of the original basic igneous rocks often occurred as sills and dykes so the various amphibolites are found as layers and bands within metamorphosed sedimentary rocks. Amphibolites are relatively common in regionally metamorphosed areas.

GRANULITES Granulites are restricted to the central parts of the world's largest continents. Granulites derived from basic rocks are composed of plagioclase feldspar (andesine or labradorite), hypersthene, diopside, and occasionally garnet (deep-red almandine), sometimes with a little quartz. The pyroxenes are derived from the breakdown of the hornblende found in the amphibolites. Owing to the abundance of pyroxene, the granulites are called **pyroxene granulites** (*see* page 161).

hornblende

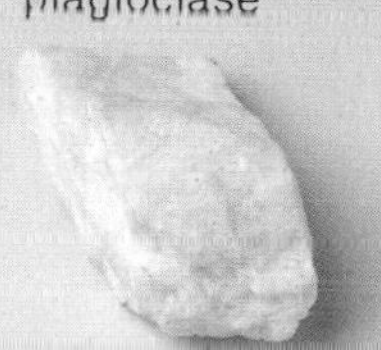

These rocks are medium- to coarse-grained, dark, granular, hard rocks, which are not usually banded or foliated. The presence of garnet gives the granulite a red speckled appearance.

Other minerals present include some brown hornblende, magnetite, ilmenite, enstatite, olivine, and the rare blue mineral sapphirine.

ECLOGITES Eclogites occur worldwide, forming isolated bodies ranging from about 1 m (2 to 3 ft) wide, to several tens of kilometres in diameter, most commonly being found within other metamorphic rocks, such as schists, gneisses, granulites, and blueschists. They can also be associated with peridotites, serpentinites, and kimberlites. Eclogites are medium- to coarse-grained rocks, usually massive and dense and having a striking greeny red speckled appearance. This coloration is due to the unique combination of the two dominant minerals, a pale-green pyroxene, omphacite, and a reddy brown garnet, pyrope. Both these minerals can occur as porphyroblasts, although the texture of eclogites is commonly granoblastic. Other minerals present in lesser amounts include another red garnet, almandine, together with rutile, kyanite, amphibole, and, in some cases, diamond. It is believed that eclogites are formed at depths of around 160 km (100 miles) below the Earth's surface.

Occurrence

When weathered all metamorphosed basic rocks are fairly easy to extract, especially those that have a schistosity or foliation. When fresh, however, they can be hard and resistant. Foliated and schistose rocks can be easily split along the foliation/schistosity by careful use of a bolster chisel.

Regionally metamorphosed basic rocks may form extensive tracts of land, depending on the distribution of the source rocks, but do not, by themselves, form large areas of high relief. Most of them are found in dyke- or sill-like bodies.

GREENSCHISTS These are found around Lake Wakatipu, and about 120 km (70 miles) north along the Haast River Valley, South Island, New Zealand; Shiojiri area, Ryoke and the Abukuma area, Japan; the northern Michigan area; Chester in Vermont and New Hampshire, USA; and the Loch Fyne area, Argyllshire, Scotland.

BLUESCHISTS These are found in the Bessi-Ino area and the Kanto Mountains of the Sanbagawa region, Honshu, Japan; the Panoche Pass area in the Diabolo Range and Ward Creek near Cazadero, about 100 km (60 miles) to the north-west, both in the Central Coast Ranges of California, USA, and New Caledonia, off the east coast of Australia.

AMPHIBOLITES These are found in the Grand Canyon area, Arizona, USA; the Lake Manapouri region, Southland area, southern South Island, New Zealand; Tanunda Creek, southern Australia, and the epidiorites of the Scottish Highlands, for example, north-west of Lochgilphead, Strathclyde, south-western Scotland.

PYROXENE GRANULITES These are found in Hartmannsdorf, Saxony, Germany; the Broken Hill district, south-central Australia; the Adirondack mountains, New York State, USA; Central Highlands area (Rangala and Gampatia), Sri Lanka; Moldanubian prov-

garnet amphibolite
hornblende schist
5cm

amphibolite
epidiorite

ince, Austria, and the extreme north-west of Scotland, such as Scourie.
ECLOGITE This may be found in the Coast Ranges in California; the western and central European Alps (Sesia-Lanzo area, north-western Italy); Nordfjord area, Norway; Glenelg in the Highland Region of western Scotland; the Sanbagawa region, Honshu, Japan, and as xenoliths in basaltic rocks at Delegate, southern New South Wales, Australia.

Mineralogy (*see* page 155)

ACTINOLITE This is a green amphibole with long prismatic crystals which are sometimes fibrous. The hardness is 5 to 6, so it may be just scratched by a penknife. There are two good cleavages, intersecting at approximately 120°, a vitreous lustre, and white streak. The fibrous habit may be diagnostic.
CHLORITE Chlorite is a green, mica-like mineral, which can also be yellow or brown. The crystals are flaky or tabular with a hardness of 2½ on cleavage planes; therefore it is easily scratched, possibly with a thumbnail. It has one perfect cleavage, a dull vitreous lustre, and white to pale green streak. The colour and cleavage are diagnostic.
EPIDOTE Epidote is a group name for several minerals which are green, greenish grey, yellowish brownish green or black. The crystals may be prismatic with striations but can be massive, fibrous, or granular. It has a hardness of 7 and does not scratch with a penknife, one perfect cleavage parallel to the length of the crystal, a vitreous lustre, and white streak. Epidote is distinguished by its colour, prismatic form and cleavage – but for the latter it could be mistaken for tourmaline.
DIOPSIDE This is a dark-green to black pyroxene which is generally paler than augite but otherwise similar.
GLAUCOPHANE Glaucophane is an amphibole, with a pale-blue, lavender-blue, dark-blue to black colour and prismatic or needle-like crystals which may even be fibrous. It has a hardness of 6, so it can just be scratched by a penknife, typical amphibole cleavage, white or blue-grey streak, and vitreous lustre.
GARNET Both pyrope and almandine are dark red so they may be difficult to tell apart in hand specimen. The well-formed crystals have a rounded shape, with twelve or twenty-four regular faces. The hardness is 7 to 7½ so they cannot be scratched with a penknife. They have no cleavage, a dull vitreous lustre, and white streak. The colour, crystal form, and hardness are diagnostic.
OMPHACITE This is a pale grass-green pyroxene. Discrete crystals are very rare; it is more often found as granular aggregates or elongate crystals. It can be scratched with a penknife (hardness 6 to 7). It has typical pyroxene cleavage, dull vitreous lustre, and white streak.
JADEITE Jadeite is a pale- to dark-green pyroxene. Discrete crystals are rare as it is usually found as granular masses. The hardness is 6 to 6½, so jadeite may just be scratched with a penknife. It has typical pyroxene cleavage, a vitreous lustre, and white streak. The colour is diagnostic but it could be confused with omphacite in hand specimen.

hypersthene

diopside

plagioclase

omphacite

garnet

pyroxene granulite

massive granulite

eclogite

5cm

Metamorphosed clay rocks

Clay rocks are particularly susceptible to both thermal and regional metamorphism, and display striking changes in their textures and mineral compositions. A shale will be taken as our example of an unmetamorphosed clay rock.

Thermal metamorphism

Thermally metamorphosed clay rocks in the aureole surrounding an igneous intrusion reflect their originally rather conservative composition and the same broad rock types and sequences are developed. The changes that occur are best described as if we imagine walking from unmetamorphosed shales towards an igneous intrusion.

SPOTTED SHALES The first visible sign of metamorphism is that the shales become spotted. These spots are grey or black and are easily visible with the naked eye or through a hand lens. At first, they are small, about 0.5 mm (0.02 in) across, but, towards the intrusion, they become bigger, often reaching 3 mm (0.1 in) in size. In some cases, the spots represent tiny concentrations of iron ore.

Individual crystals cannot be recognized in these rocks, but they have become a little coarser grained and contain some microscopic chlorite flakes. Original sedimentary structures and fossils are still recognizable, however.

Moving nearer to the intrusion, more changes are found. The shale is harder, and any structures and fossils become obliterated. Biotite is often the first new mineral to develop, and is recognized as dark-brown flakes forming small porphyroblasts.

CHIASTOLITE SLATE Next to form are andalusite and cordierite, sometimes forming porphyroblasts. Depending on the original composition of the rock, one or both may grow. Usually, if both are present, andalusite develops first but always after biotite. The variety of andalusite called **chiastolite** is easily recognized as it has very fine-grained dark inclusions arranged in a characteristic cross shape. Chiastolite may form quite large, abundant porphyroblasts, and the rock is then called a **chiastolite slate** or **chiastolite hornfels**. The rock also contains quartz, biotite, muscovite, andalusite, and/or cordierite. Again, although it is not obvious to the naked eye, a further increase in grain size has occurred.

HORNFELS Close to the igneous intrusion, and possibly right up to the contact, the clay rock is seen to be metamorphosed to a hard, splintery, dark, fine- to medium-grained hornfels, often having a pale grey to white, bleached surface appearance. Crystals may be visible, and it may be possible to recognize the characteristic granoblastic texture. There is some variation in the minerals which develop in these hornfelses. Andalusite and cordierite are still present, often forming relatively large porphyroblasts. These are usually the commonest minerals and the rock is termed an **andalusite-cordierite hornfels**. Orthoclase may

andalusite

cross section through
chiastolite

cordierite

'spotted' rocks

chiastolite slate

5cm

hornfels with
andalusite and
chiastolite

banded hornfels

hornfels

develop instead of muscovite, but this would not be apparent in hand specimen. Occasionally, microcline may be present. Tourmaline may be found as an accessory mineral but, again, it is doubtful whether this would be recognized in hand specimen. In similar hornfelses which lack silica, quartz may not be present; instead, spinel and corundum may be there.

Within clay hornfelses, very near to the igneous contact, hypersthene may be found developed to such an extent that the rocks are called **pyroxene hornfelses** and contain combinations of hypersthene, cordierite, spinel, corundum, and orthoclase; plagioclase may also be present. It may be difficult to distinguish these pyroxene hornfelses from those formed from the thermal metamorphism of basic igneous rocks although the basic hornfelses are often darker. Their occurrence in the field, however, usually makes identification relatively easy.

Another mineral often found in clay hornfelses very close to an igneous intrusion is sillimanite, which forms at the expense of andalusite. Other minerals which are rather infrequently found in these hornfelses include kyanite, staurolite, and almandine garnet, but they are more typically found in regionally metamorphosed clay rocks.

Occurrence

These hornfelses are usually hard and splintery and can be difficult to extract when fresh. Take care to protect your eyes from flying sharp fragments when hammering these rocks. Spotted shales and chiastolite slates can be collected using a bolster chisel, because the rocks are readily split. Certain metamorphosed clay rocks may develop particularly large chiastolite crystals several centimetres long but, to preserve the outcrop, you should not try to collect them.

Metamorphic aureoles are limited in their extent, so that thermally metamorphosed clay rocks do not have much influence on the landscape. The original rocks were soft and easily eroded, and did not greatly affect the landscape, tending to produce rather flat land. The hornfelses nearest the intrusion are hard and resistant, however, and they may produce localized areas of high ground with angular rock exposures; they can even be harder than the igneous rock itself.

Thermally metamorphosed clay rocks are found in the following areas: the Comrie district, Perthshire, Scotland; the Glen Cova area, Angus, Scotland; the Skiddaw area, Cumbria, in the English Lake District; the Marysville area, Montana, USA; the Donegal area, north-western Ireland; Cliffe Hill Quarry in the Markfield area, Leicestershire, England; the Cascade Valley, Westland, New Zealand; certain areas in the Santa Rosa range, northern Nevada, USA, and Leuchtenberg, Steinach area, north-east Bavaria, Germany.

Regional metamorphism

Of all the different rock types, clay rocks are generally the first to respond to regional metamorphism. Taking shale again as the starting point, the first change is its conversion into a slate.

black slate

green pyritiferous slate

purple slate

crenulated pyritiferous slate

black pyritiferous slate

weathered pyritiferous slate

5cm

phyllite

SLATES These can be black and shades of brown, green, and blue. They are extremely fine grained but original sedimentary structures may still be visible. Slates possess an excellent foliation or slaty cleavage, and can be split readily along the cleavage planes which run through the rock using a bolster chisel. The slaty cleavage is caused by the abundance of minute flakes of mica which all lie roughly parallel to one another. The micas are mainly chlorite and muscovite flakes which have recrystallized from the clay minerals present in the unmetamorphosed shale. The black colour of some slates is due to the presence of tiny carbon particles. Certain slates may contain quite large, well-formed cubes of pyrite and are known as **pyritiferous slates**. On the surface exposures, these rocks develop cavities often surrounded by a rusty brown deposit, where the cubes have weathered away (*see* page 165).

PHYLLITE At a constant pressure, but with a slight rise in temperature, up to about 300 to 350°C (570 to 660°F), a distinctive rock called a **phyllite** is formed. Phyllites are slightly coarser grained than slates, although this may not be obvious in hand specimen. Phyllites are rich in muscovite, and particularly chlorite, which gives them their common pale-green colour. They have a well-developed cleavage or foliation (schistosity), due to the parallel alignment of all the mica flakes within the rock (*see* page 165). As these flakes are somewhat coarser than those in the slates, their tiny reflecting cleavage surfaces give phyllites a characteristic lustrous, silky sheen. The cleavage planes are commonly wrinkled and rumpled with irregular surfaces. Certain phyllites may possess thin, quartz-rich bands which are parallel to the schistosity. Common minerals are chlorite, muscovite, some quartz and iron ore, possibly with a little biotite.

SCHISTS Further rises in temperature and pressure (and metamorphic grade) produce these common metamorphic rocks. The various types of schist are named after their dominant minerals. The grain size is coarser than those of slates and phyllites. As their name implies, these rocks possess a foliation or schistosity. This characteristic feature of schists is clearly visible.

Chlorite schists are low-grade schists, and are green-grey with a good schistosity. Chlorite is commonly found as porphyroblasts, as is a similar blue-green mineral, **chloritoid**. Chlorite schists also contain quartz, muscovite, and epidote, with possibly some plagioclase feldspar (albite). Such rocks are often found associated with phyllites.

Slightly higher-grade schists are the **mica schists** such as biotite schists and muscovite schists. Muscovite schists are pale, lustrous rocks with a good schistosity and are relatively coarse grained. Some quartz is often present, together with a little chlorite, biotite, and deep-red garnet. Likewise, biotite schists can be coarse grained but are much darker. They possess a good schistosity and are frequently banded. Some chlorite and muscovite may be present but it is biotite that usually develops. Quartz and feldspar are also present, located within the paler bands; feldspar may also form porphyroblasts. Biotite schists may contain deep red almandine garnets as prominent porphyroblasts. These distinctive dark rocks

chlorite
muscovite
biotite
garnet
staurolite
kyanite
sillimanite
quartz
plagioclase
schist
5cm

with red patches are called **garnet-mica schists** and, in addition, they contain quartz, muscovite, and plagioclase feldspar. All the chlorite has now disappeared. Schists containing abundant garnet are formed at slightly higher grades of metamorphism than the mica schists. Garnet-mica schists and mica schists are often folded and crenulated.

Higher-grade schists are the **staurolite** and **kyanite schists**. These still possess a good schistosity and are banded. They are usually dark, fairly coarse-grained rocks, having porphyroblasts of garnet, staurolite, and kyanite. The staurolite is often randomly arranged with respect to the schistosity. The common minerals in these schists are quartz, biotite, muscovite, garnet, kyanite, staurolite, and plagioclase (oligoclase), possibly with a little orthoclase. Generally, staurolite develops before kyanite.

Occasionally, at slightly higher grades of metamorphism, sillimanite replaces kyanite to form **sillimanite schists**. While garnet, staurolite, and kyanite form distinctive porphyroblasts, sillimanite crystals are often delicate and needle-like.

Accessory minerals in these schists are apatite, tourmaline, zircon, and black iron ore.

GNEISS (*see* p. 171) These rocks are formed at even higher grades. They are usually medium to coarse grained and pink-grey in colour. The banding, described as gneissose banding, is well defined. The dark bands consist of mica and hornblende and the light bands of quartz and feldspar. The bands vary in thickness from millimetres to several centimetres, and can be either straight or folded. Gneisses are well foliated and, in the dark layers, the hornblende crystals may be lineated. Granoblastic textures are developed in the pale quartz- and feldspar-rich bands. Common minerals are quartz, feldspar, biotite, hornblende, kyanite, and sillimanite, with possibly some garnet. Sometimes lens-shaped porphyroblasts of feldspar are developed which measure over 1 cm ($\frac{1}{2}$ in) across and the rock is called an **augen gneiss**. A further increase in metamorphic grade may promote localized melting of the gneisses to produce rocks called migmatites.

Occurrence

Regionally metamorphosed clay rocks are relatively easy to extract, because of their foliation, and can be readily split with a bolster chisel. Gneisses do not split so easily and, when freshly exposed, are hard and resistant. Slates, phyllites, and schists weather fairly readily, and can become quite soft and crumbly, sometimes acquiring a rusty brown colour.

These rocks form extensive tracts of land of variable relief. They are frequently found in many of the world's mountain chains although the high relief is because they have been folded and thrust upwards rather than being due to their hardness. The more resistant gneisses can, however, form upland tracts of land.

Areas where such rocks are found include Bethesda and Blaenau Ffestiniog, North Wales (slates); parts of Scotland (schists, gneisses); central Maine and eastern Vermont, USA (slates); Orange, in southern

chlorite schist
mica schist
garnet-mica schist
5cm
staurolite schist
biotite kyanite schist
mica schist

Connecticut, USA (phyllites); the Appalachians, eastern America (schists); Windham in Maine, USA (staurolite-schist), and the European Alps (schists, gneisses).

Mineralogy (*see* page 167)

ANDALUSITE The prismatic crystals have a square cross-section and can be white, grey, pink, brown and green. The two cleavages are at roughly 90° but, in hand specimen, only one cleavage may be seen. They cannot be scratched with a penknife (hardness 7½), give a white streak, and have a vitreous lustre. The shape and hardness of andalusite are diagnostic. The variety chiastolite contains dark inclusions in a cross-like arrangement.

CORDIERITE This is more common as massive or irregular grains; they are usually pale to dark blue in colour but can be grey. There is one poor cleavage and the hardness is 7 so it cannot be scratched by a penknife. It also has a vitreous lustre and a white streak. The blue colour is diagnostic but, when it is grey and granular, it appears very similar to quartz and can only be accurately distinguished under a microscope.

CORUNDUM The crystals are tabular, prismatic, or barrel shaped. The colour can be white, yellow-brown, yellow, blue (sapphire) or red (ruby). There is no cleavage and corundum is so hard (9) that it can only be scratched by diamond. It has a white streak and a brilliant vitreous lustre. The diagnostic features are the crystal form and, in particular, the hardness.

STAUROLITE The crystals are usually prismatic and reddish brown or dark brown in colour with one cleavage. It cannot be scratched with a penknife (hardness 7). It has a dull but vitreous lustre and a grey streak.

KYANITE Closely related to andalusite and sillimanite in composition, kyanite has flat, bladed crystals of a distinctive pale-blue colour, but it can be white, grey or green, the coloration often being patchy. It has two good cleavages. The hardness is variable, between 4 and 7, and the crystals can be easily scratched along, but not across, their length. Kyanite has a vitreous to pearly lustre and white streak. The colour and the bladed form are characteristic.

SILLIMANITE Closely related to andalusite and kyanite, sillimanite generally has prismatic crystals but, when it is fibrous, it is known as **fibrolite**. It is white, pale brown, or pale green in colour with one good cleavage. The hardness is 7, so it is not easily scratched. It has a vitreous lustre, white streak, and characteristic fibrous form. This does not distinguish sillimanite from other fibrous minerals, however, and it can only be identified accurately under a microscope.

gneiss
gneiss
garnet hornblende gneiss
hornblende biotite gneiss
gneiss
5cm
garnet sillimanite gneiss
augen gneiss
gneiss

Metamorphosed sandstones

Thermal metamorphism

QUARTZ SANDSTONES These fairly pure sandstones are relatively resistant to low- and medium-grade thermal metamorphism. They often contain some clay minerals, however, which recrystallize into small muscovite flakes lying between the quartz grains although, in hand specimen, such minor changes may not be visible. At higher grades, the quartz grains simply undergo recrystallization to form a **quartzite** with a granoblastic texture. These extremely hard, creamy white rocks are more accurately termed **metaquartzites** or **quartz hornfelses**, to avoid confusion with sedimentary quartzites. Both types are generally structureless, but the presence of small glistening muscovite flakes may serve to distinguish metaquartzites which are also often heavier and harder.

ARKOSE AND FELDSPATHIC SANDSTONE High-grade thermal metamorphism of these sandstones produces massive, hard, splintery, pale hornfelses. Again these rocks become recrystallized into granoblastic aggregates of quartz and feldspar. Any original clay minerals have been converted to biotite, while the alkali and plagioclase feldspars may undergo alteration to muscovite.

GREYWACKES These all contain quartz, feldspar, mica, and clay minerals but the often abundant rock fragments frequently show a range of composition so that **metagreywackes** themselves can be somewhat variable. Low-grade thermal metamorphism promotes recrystallization of the fine-grained matrix, originally of quartz and clay minerals, to a granoblastic aggregate of slightly coarser quartz and randomly arranged mica flakes, with possibly some fine amphibole. The larger crystals and rock fragments remain unaltered, and it may be difficult to see any metamorphic changes in hand specimens, which often appear as dark, massive rocks.

Higher grades of thermal metamorphism promote recrystallization of large crystals and rock fragments to produce medium-grained, dark hornfelses. High-grade metagreywackes are dark, tough, even grained, and 'igneous-looking'. Depending on the original rock composition, a variety of minerals are developed including biotite, andalusite, and cordierite (as porphyroblasts), hypersthene, quartz, feldspar (usually orthoclase), garnet, and staurolite. If the hornfelses have a reduced silica content, corundum and spinel may be found.

Regional metamorphism

QUARTZ SANDSTONES These again become regionally metamorphosed to **quartzites**, and, in hand specimen, they are difficult to distinguish from quartzites produced by thermal metamorphism. In some cases, a high-grade regionally metamorphosed quartzite may develop a foliation but, again, this can be difficult to recognize in hand specimen. To distinguish accurately the two types of quartzite, the field

banded hornfels

quartzite

5cm

hornfels

occurrence is valuable: one found near to an igneous intrusion is likely to have been thermally metamorphosed while a quartzite from an area of folded rocks is likely to have been formed by regional metamorphism.

ARKOSE AND FELDSPATHIC SANDSTONE These are generally converted to pale, foliated rocks composed predominantly of recrystallized granoblastic quartz and feldspar (orthoclase and plagioclase), the latter sometimes occurring as porphyroblasts. Such rocks may be termed **quartz schists**, and the crystals are visible to the naked eye. Any original clay minerals become recrystallized into mica, generally muscovite, although biotite may also be present. These micas tend to lie parallel to one another and define a foliation. Muscovite may also be derived from the alteration of the feldspar orthoclase. Some microcline may also be present.

GREYWACKES Low-grade metamorphism gives rise to zeolites; at this stage, no discernible change in the rocks is visible. With a slight increase in metamorphic grade, the minerals **prehnite** and **pumpellyite** are formed and are associated with quartz, plagioclase feldspar (albite), chlorite and possibly sphene. The amphibole, actinolite, and some muscovite may form at this stage. In hand specimen, the metagreywackes appear unmetamorphosed.

Further rises in metamorphic grade are accompanied by a variably developed foliation. With a rise in pressure only, **blueschists** are formed once again (these can be derived from both sedimentary and basic igneous rocks), composed mostly of quartz, jadeite, glaucophane, muscovite, plagioclase (albite), some chlorite, and a mineral called **lawsonite**. In hand specimen, when fresh, the fine- to medium-grained rock is dark blue-grey and foliated; small glistening flakes of mica are visible on the exposed foliation surfaces.

Higher metamorphic grades produce foliated schists and banded gneisses composed mostly of quartz, plagioclase, microcline, biotite, and muscovite, occasionally with hornblende, the micas defining the foliation. Such rocks usually possess well-defined light and dark bands.

At even higher grades, the metamorphic rocks tend to lose their foliation and take on a granular appearance and granoblastic texture. Quartz, plagioclase, orthoclase, and biotite are the main mineral constituents of these rocks, which are often called **granulites**.

Occurrence

Metamorphosed sandstones may form areas of high land but their distribution is obviously dependent on that of the original sandstones. Apart from the extremely hard and durable quartzites, metamorphosed sandstones are relatively easy to extract with a hammer and bolster chisel.

Those rocks produced by thermal metamorphism occur north of Comrie, Perthshire, southern Highlands of Scotland; the Oslo region, Norway and Hessen in Germany. Examples produced by regional metamorphism include the north-western Adirondacks, New York State, USA; the Central Coast Ranges (Diabolo Range), USA; parts of the Scottish Highlands; and Anglesey in North Wales.

metamorphosed grit

foliated quartzite

blueschist

quartzo-feldspathic schist

5cm

granulite

gneiss

gneiss

Metamorphosed limestones

Limestones are readily metamorphosed. A useful diagnostic feature of such rocks, especially those with appreciable amounts of calcite, is their vigorous effervescence with dilute hydrochloric acid.

Thermal metamorphism

PURE LIMESTONE Thermal metamorphism of rocks composed almost entirely of calcite involves the recrystallization of the calcite into a granoblastic aggregate resulting in an even-grained, white **marble**. The surface appearance often resembles white sugar.

DOLOMITE Not all limestones are pure, however, and can contain dolomite, silica (as sand grains or chert fragments), and land-derived material (usually as clay minerals). At low grades, dolomites and dolomitic limestones simply recrystallize into pale **dolomitic marbles**, which, in most respects, appear very similar to marbles. Dolomitic marbles may not react as strongly with dilute hydrochloric acid.

In some cases, the extremely soft mineral, talc, may form, as may epidote, which imparts a pale greenish tinge to the rock. These are more common, however, in metamorphosed limestones which originally contained some impurities.

At higher grades, dolomite breaks down, forming calcite, a mineral called **periclase**, and carbon dioxide which is lost to the atmosphere. These rocks are termed **periclase marbles**. Periclase is unstable, and atmospheric water may be added to its atomic structure to form **brucite**. Any metacarbonate with appreciable amounts of brucite is called a **brucite marble** or **predazzite**, named after Predazzo in Italy.

SILICEOUS LIMESTONES At low grades of thermal metamorphism, only the calcite recrystallizes, to produce quartz marbles. With a temperature increase, recrystallization of some of the calcite and quartz produces the mineral **wollastonite** and carbon dioxide. Calcite is usually more abundant than quartz in siliceous limestones so that which is not used up in producing wollastonite is recrystallized into a granoblastic aggregate which may be visible with the naked eye or through a hand lens. The rock is known as **wollastonite marble**.

OTHER LIMESTONES Some limestones are a mixture of calcite, dolomite, silica in the form of quartz grains or chert fragments, and possibly even some clay minerals. At low grades, only the calcite and dolomite in such rocks would recrystallize, although the clay minerals may recrystallize to produce some mica flakes. With the onset of high-grade thermal metamorphism, the dolomite and silica components recrystallize to produce more calcite and a variety of olivine called **forsterite**. The resulting pale grey, medium- to coarse-grained crystalline rock is known as a **forsterite marble**. Like periclase, forsterite may become hydrated to form the green fibrous, flaky mineral, serpentine, when the rock is called a serpentine marble, or **ophicalcite**.

Ophicalcites have a creamy white weathered surface but, when fresh,

calcite
brucite
tremolite
marble
brucite marble
marble
wollastonite marble
diopside marble
forsterite marble
5cm
tremolite marble

they are often delicately tinted with swirls of pale to vivid green. The minerals talc, **tremolite** (a calcium-magnesium-rich amphibole), diopside, and **grossular**, a pale-green garnet containing calcium and aluminium are also developed.

If the amount of clay minerals in the original limestone is quite high, their metamorphic equivalents may be composed almost entirely of calcium-rich silicate minerals rather than calcite. Such minerals include tremolite, diopside, grossular, calcium-rich plagioclase (anorthite, bytownite and labradorite), a dark-green mineral called **idocrase** or **vesuvianite**, wollastonite, varieties of epidote, **scapolite**, and even a little quartz. These rocks are called **calc-silicate hornfelses.**

Usually, thermal metamorphism involves no overall change in the composition of the rocks, apart from the addition or loss of water and carbon dioxide. Sometimes, changes do occur, usually by addition of elements such as silicon, aluminium, iron, magnesium, chlorine, fluorine, and boron. These are transported by fluids, associated with the igneous intrusion, which pervade the country rock. Limestones are particularly susceptible to this process, called **metasomatism**. Such modified hornfelses are termed **skarns** and are found within quite narrow, well-defined zones (several metres wide) immediately adjacent to the igneous body. Skarns can be fine- to coarse-grained rocks, and may show a range of colours; excellent specimens of minerals can often be obtained from these rocks.

Regional metamorphism

PURE LIMESTONES AND DOLOMITES These become recrystallized to form marbles, possibly with the development of some foliation.

More massive marbles may only be distinguished from thermal metamorphic marbles if their field relationships are known.

SILICEOUS LIMESTONES AND DOLOMITES These may simply recrystallize to form poorly foliated quartz-calcite or quartz-dolomite rocks but, if water is present, the dolomite may break down to form talc or, at slightly higher grades, tremolite. At even higher grades diopside and then forsterite are formed in sequence. Some epidote and grossular may also be developed together with the formation of wollastonite at high grades. These rocks may be foliated (**calcareous** or **calc-silicate schists**) but can be massive. If clay impurities are present in the original limestone, micas (muscovite, biotite, or phlogopite) may be generated to give well-foliated schistose metacarbonates. This foliation may be lost at higher grades, while well-banded **calc-silicate gneisses** may also form. The presence of calcite, tremolite, diopside, epidote, and grossular serve to make many regionally metamorphosed calc-silicate rocks unfoliated and granular, so that they appear rather similar to their calc-silicate hornfelsic counterparts. Only field relationships accurately distinguish them. Scapolite is a relatively common mineral in regionally metamorphosed carbonate rocks.

OTHER LIMESTONES At high metamorphic grades, impure

idocrase
grossular garnet
scapolite
5cm
talc
ophicalcite
banded ophicalcite
garnet skarn
garnet idocrase skarn

limestones (with aluminium and silica) may be converted into rocks containing abundant calcium-rich plagioclase feldspar, in addition to amphibole or pyroxene.

Occurrence

In general, metamorphosed limestones are not resistant rocks and weather fairly quickly to show creamy white weathered surfaces. They can be extracted relatively easily. By themselves, they do not tend to influence the landscape to any great extent, but they can occur in association with harder rocks forming upland areas.

Areas of thermally metamorphosed limestones include the Beinn an Dubhaich area, southern Skye, Inner Hebrides off western Scotland; the Three Rivers area, Sierra Nevada, California, USA; Predazzo in Italy; Ben Bullen, New South Wales, Australia, and localized areas around the granites of Cornwall and Devon in south-west England.

Areas of regionally metamorphosed limestones include Strathspey in Scotland; the Castleton area, Vermont and areas about 50 km (30 miles) east, and the north-western Adirondacks, New York State.

Essential minerals (*see* pages 175 and 177)

BRUCITE The crystals are tabular, platy, foliated, massive, or fibrous with one perfect cleavage. They may just be scratched with a fingernail (hardness 2½) and are white or pale grey, blue, or green in colour with a white streak. The cleavage planes have a pearly lustre. The cleavage, low hardness, and foliated form are diagnostic.

WOLLASTONITE This occurs as tabular or prismatic crystals or in radiating fibrous masses. The three cleavages may be difficult to recognize. It can be easily scratched with a penknife (hardness 4 to 5). Wollastonite is white to grey in colour with a white streak and vitreous to silky lustre. Similar to other white fibrous silicates.

TREMOLITE It is similar to actinolite except that tremolite is white to very pale green in colour.

GROSSULAR This garnet is essentially similar to pyrope and almandine except that grossular is pale green.

IDOCRASE This is prismatic or massive, often striated, and may just be scratched with a penknife (hardness 6 to 7). It is usually green or brown with a white streak, vitreous lustre, and poor cleavage. The diagnostic features of the crystals are their prismatic form and striations, but massive aggregates may be mistaken for garnet or epidote.

TALC Talc is often found as flaky masses or compact aggregates. This mineral is extremely soft (hardness 1), pale green, white, or grey with a white streak and one perfect cleavage. It generally displays a dull lustre. Its colour, softness, and soapy feel are diagnostic.

SCAPOLITE Scapolite occurs as prismatic crystals but is more generally massive or granular. It has two cleavages and can be scratched with a penknife (hardness 5 to 6). It varies from white to bluish grey in colour with a white streak and vitreous lustre. The massive appearance, colour, and cleavage are all diagnostic for scapolite.

calc-silicate hornfels

marble with garnet

calcareous schist

banded marble

5cm

grossular idocrase marble

marble with mica

Bibliography

Berry, L. G. and Mason, B. *Mineralogy*. London: Bailey and Swinfen, 1959. New York: W. H. Freeman, 1959.

Cox, K. G., Price, N. B. and Harte, B. *An Introduction to the Practical Study of Crystals, Minerals and Rocks*. Revised edition. Maidenhead, England: McGraw Hill, 1974.

Dietrich, R. V. and Skinner, B. J. *Rocks and Rock Minerals*. Chichester, England: John Wiley, 1979.

Ehlers, E. G. and Blatt, H. *Petrology – Igneous, Sedimentary and Metamorphic*. New York and Oxford: W. H. Freeman, 1982.

Francis, P. *Volcanoes*. Harmondsworth, England: Penguin, 1976.

Friedman, G. M. and Sanders, J. E. *Principles of Sedimentology*. Chichester, England: John Wiley, 1978.

Hatch, F. H., Wells, A. K. and Wells, M. K. *Petrology of the Igneous Rocks*. 13th ed. London: Thomas Murby, 1972.

Holmes, A. *Principles of Physical Geology*. Revised edition. Walton-on-Thames, England: Nelson, 1965.

Mason, R. *Petrology of the Metamorphic Rocks*. London: George Allen and Unwin, 1978.

Rutley, F. *Rutley's Elements of Mineralogy*. 26th edition. Edited by H. H. Read. London: Thomas Murby, 1978.

Selley, R. C. *Ancient Sedimentary Environments and their Sub-Surface Diagnosis*. 2nd edition. London: Chapman and Hall, 1978.

Tucker, M. E. *Sedimentary Petrology – an Introduction*. Oxford: Blackwell Scientific, 1981.

Whitten, D. G. A. and Brooks, J. R. V. *The Dictionary of Geology*. Harmondsworth, England: Penguin, 1972. Harmondsworth, England: Allen Lane, 1978.

Glossary

Acid Describes igneous rocks having more than (>) 10% visible quartz, and which are rich in alkali feldspars.

Amphiboles Large, complex group of chain silicates, rich in calcium, iron, magnesium and aluminium, often forming fibrous or acicular crystals, e.g. hornblende.

Amygdale Vesicle which has been later infilled by secondary minerals, namely zeolites, calcite or quartz.

Anhedral Term applied to crystals in igneous rocks showing no development of regular crystal faces.

Augen Lenticular, 'eye-shaped', mineral concentration, containing coarse feldspar or quartz crystals; often found in gneisses.

Aureole Zone around an igneous intrusion in which the country rocks have undergone thermal metamorphism.

Basic Describes igneous rocks having no quartz but possessing olivine, pyroxenes and calcium-rich plagioclase feldspars.

Batholith Any large, intrusive body of igneous rock, with no observable bottom, frequently composed of granitic rocks.

Bedding (Usually) a plane representing a surface parallel to the original surface of deposition.

Bioclast Any clast that is a fragment of organic skeletal material.

Bitterns Extremely soluble chemicals that are precipitated only in the final stages of sea water evaporation.

Cement Material binding sedimentary grains together; usually crystalline.

Clastic Consisting of fragments which have been individually moved from their places of origin.

Contact (thermal) metamorphism Metamorphism produced by increased temperature alone, and associated with igneous intrusions.

Country rock Any rock surrounding and intruded by an intrusive igneous rock.

Crystal Solid with a definite chemical composition and an ordered atomic structure, with naturally formed plane faces.

Diagenesis Those physical and chemical changes occurring in a sediment after deposition, and at relatively low temperatures.

Dip Maximum angle that a bedding plane makes with a horizontal plane.

Dyke Vertical or oblique, sheet-like, hypabyssal intrusion of igneous rock which is discordant, and of variable width.

Euhedral Term applied to crystals in igneous rocks showing well-developed, regular crystal faces.

Eustatic Describes a change in sea level that is worldwide.

Fabric Term encompassing *texture* (relationship between grains in a rock), and *structure* (relations between grain aggregates, such as bedding or folding).

Facies Combination of features such as sedimentary rock type fossil content, etc., characterizing a sediment formed in a particular environment of deposition.

Fault Fracture in a rock where observable movement has occurred on one side relative to the other.

Feldspars Framework silicates rich in aluminium, potassium, calcium and sodium; the most important group of rock-forming silicates.

Felsic General term applied to light-coloured silicate minerals, such as feldspar and quartz.

Flow banding Structure frequently developed in acidic lavas; represents slight compositional differences or gas bubble concentrations.

Folding Change in the amount of dip of a bedding plane.

Foliation Fabric frequently developed in regional metamorphic rocks, where platy minerals develop a parallel orientation.

Fracture Applied to rocks and minerals, fractures are breakages not paralleling cleavage directions or recognizable structural planes.

Grade Term providing a relative measure of the intensity of metamorphism in an area.

Hypabyssal intrusion Shallowly located igneous intrusion formed by magma which almost reached the Earth's surface, e.g. dykes, sills.

Hypermelanic Describes igneous rocks consisting almost entirely of dark (mafic) minerals.

Igneous rock Rock believed to have formed from the solidification of magma.

Joint Vertical, inclined or horizontal rock fracture produced by folding or cooling; no movement is involved.

Leucocratic Describes igneous rocks rich in light-coloured (felsic) minerals, the rock itself being pale.

Lineation Parallel orientation of elongate minerals on a planar surface in a regional metamorphic rock.

Mafic General term applied to iron- and magnesium-rich, dark silicate minerals, e.g. pyroxenes and amphiboles.

Magma Molten rock with dissolved gases formed at depth, by melting; solidifies to form igneous rocks.

Massive Term implying that a rock or mineral either is heavy or has no recognizable features within it.

Matrix Any fine-grained material that occurs between grains or clasts in a sediment.

Melanocratic Describes igneous rocks consisting mainly of dark (mafic) minerals, the rock itself being dark.

Mesocratic (mesotype) Applied to igneous rocks with approximately equal quantities of light (felsic) and dark (mafic) minerals.

Metamorphic rocks Rock derived from pre-existing rocks by alteration to their mineralogy, structure and composition, caused by temperature and pressure changes within the Earth.

Microlite Extremely small, often prismatic crystal found in glassy rocks, and identified only with powerful microscopes.

Oil Trap Combination of factors which allow oil to accumulate by preventing further movement towards the surface.

Permeable Can transmit fluids.

Phenocryst Any relatively large visible crystal surrounded by much smaller crystals in igneous rocks.

Pluton Generalized term for any large body of intrusive igneous rock, irrespective of its shape.

Poikiloblast Term applied to any relatively large crystal enclosing smaller crystals, in metamorphic rocks.

Porphyroblast Any relatively large visible crystal surrounded by much smaller crystals in a metamorphic rock.

Porphyry Term describing any igneous rock having large isolated crystals surrounded by much finer crystals.

Pyroclastic Term applied to any fragmental material produced by explosive volcanic activity, excluding lava.

Pyroxenes Varied group of chain silicates, rich in magnesium, iron and calcium, e.g. augite, hypersthene and diopside.

Regional Metamorphism Metamorphism produced by increases in temperature and pressure generated on a regional scale.

Sabkha Arabian term for the broad saline flats formed along coastlines and inland which are only occasionally flooded.

Salt Dome Mass of salt rising through overlying rocks by virtue of its lower density.

Salts Whole class of chemical compounds which, when referred to geologically, describe those which precipitate as a result of sea water evaporation.

Schistosity Foliation developed in the medium – coarse-grained regional metamorphic rocks called schists.

Sedimentary Cycle Sequence corresponding to a sedimentary rhythm, resulting in a particular sequence of sediments.

Sedimentary Rocks Rocks produced from the erosion of pre-existing rocks, precipitated by organisms or crystallizing from fluids at surface temperatures.

Sill More or less horizontal, sheet-like, hypabyssal intrusion of igneous rocks which is concordant.

Spherulite Spherical aggregate 1 cm (½ in) of fine, needle-like crystals, arranged radially, resulting from the devitrification of glass.

Strike Horizontal line on a bedding plane, at right angles to the direction of dip.

Subhedral Term applied to crystals in igneous rocks showing some vestiges of regular crystal faces.

Supratidal Shore zone immediately above the high-tide line, moistened by spray and subject to only very high tides.

Ultrabasic Igneous rocks composed predominantly of dark (mafic) minerals, with no quartz and little plagioclase feldspar.

Ultramafic Describes igneous rocks composed entirely of dark (mafic) minerals, with no quartz or feldspar.

Vein Narrow, broadly sheet-like mineral body penetrating rocks; either of igneous origin or represent fractures infilled by minerals deposited by solutions, e.g. quartz, calcite.

Vesicle Small ellipsoidal or spherical cavity, originally a gas bubble, found in igneous rocks.

Xenolith Fragment of country rock which has become incorporated and preserved within an igneous rock.

Zeolites Large group of framework silicates containing calcium, potassium, sodium, barium and water; frequently found within amygdales.

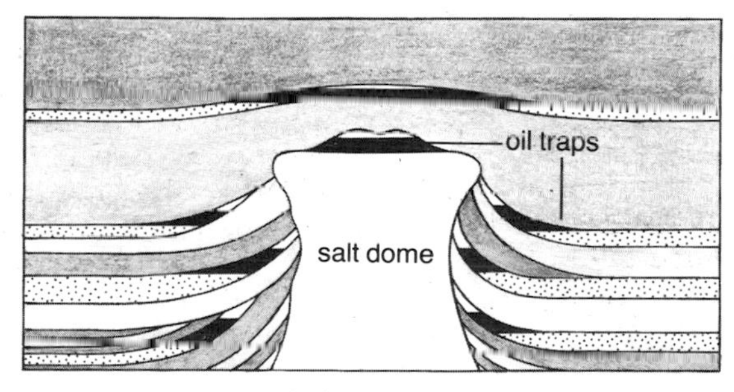

Salt and domes and oil traps *(see page 142)*
Salt domes push up through the strata forming an impermeable barrier. Because of this they are frequently associated with oil traps. The oil migrates through porous rocks, joints, and bedding planes, and then accumulates in beds adjacent to the salt when it can rise no further.

Textures found in igneous and metamorphic rocks

The following textures are commonly found in igneous and metamorphic rocks and are mentioned throughout this book. The reader is referred to the area where that particular texture is discussed in most detail.

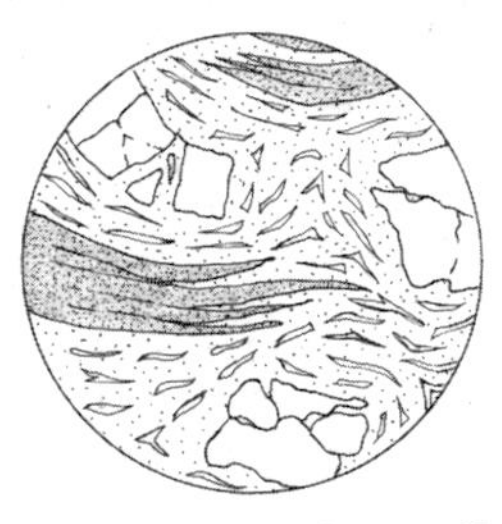

Flattened ash flow tuff *(see page 78) This rock has a distinctive streaky appearance.*

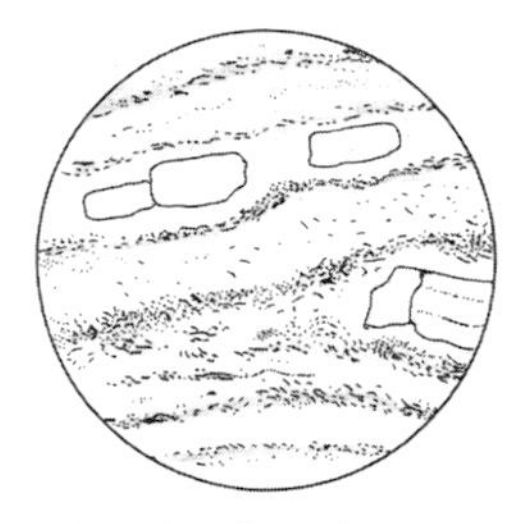

Flow banding *(see page 72) Rhyolites display this structure which is caused by slight change in grain size and colour.*

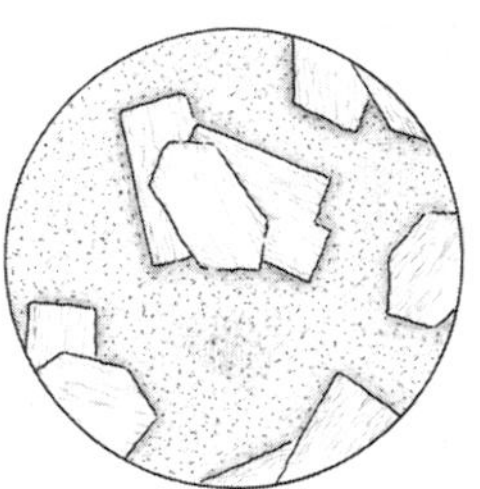

Glomeroporphyritic texture *(see page 106) A cluster of phenocrysts may, at first sight, appear as one large crystal.*

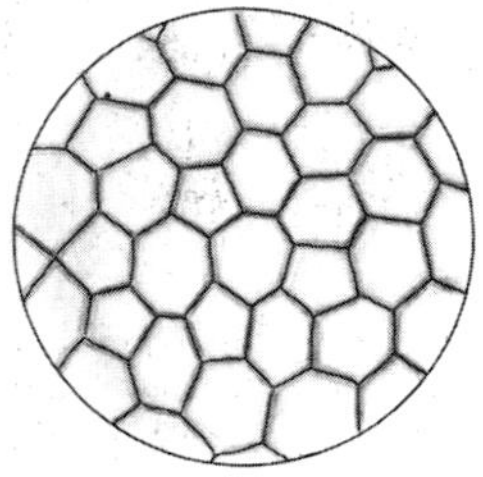

Granoblastic texture *(see page 152) An equigranular texture found in metamorphic rocks such as basic hornfels and eclogite.*

Granophyric texture *(see page 70) Complex intergrowth of quartz and feldspar crystals, similar to graphic texture.*

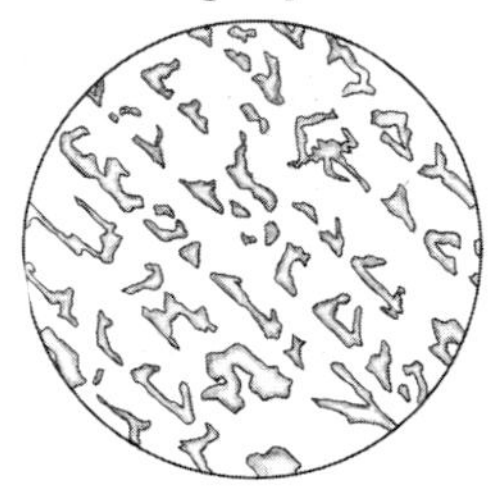

Graphic texture *(see page 66) An intergrowth of quartz and feldspar crystals found in pegmatites and granitic rocks.*

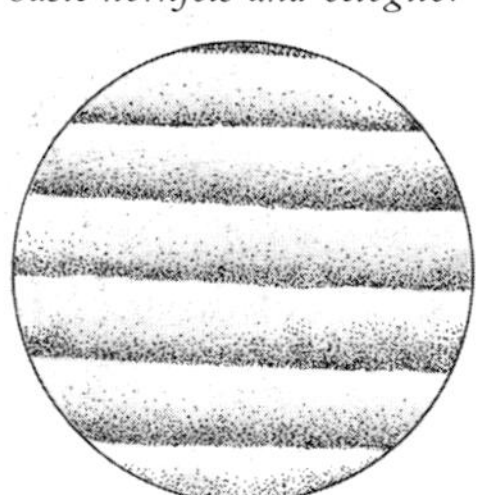

Layered gabbro *(see page 98) Slow cooling of the rock results in minerals crystallizing out in a particular order creating a banded appearance.*

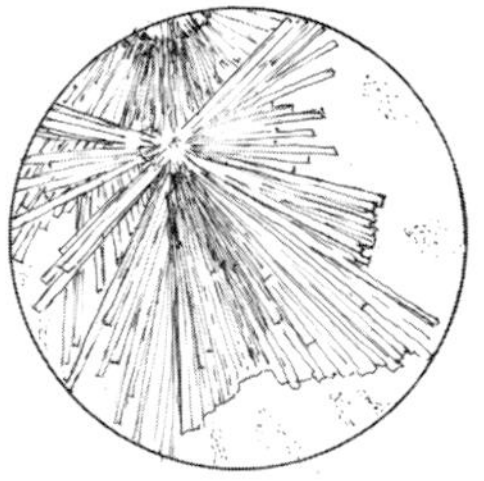

Luxullianite *(see page 82) The radial arrangement of tourmaline crystals in this rock is not always clearly visible to the naked eye.*

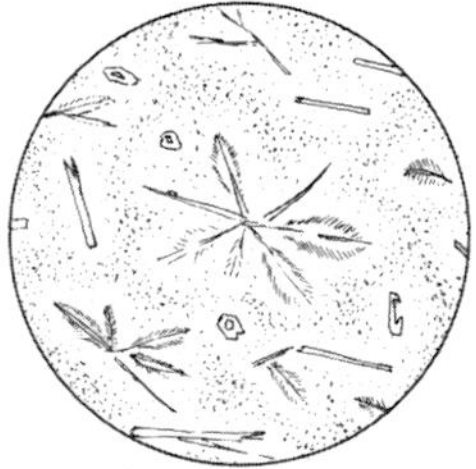

Microlites in glass *(see page 74) These are minute fibrous crystals which gradually form and replace glass in rapidly cooled rhyolitic rock.*

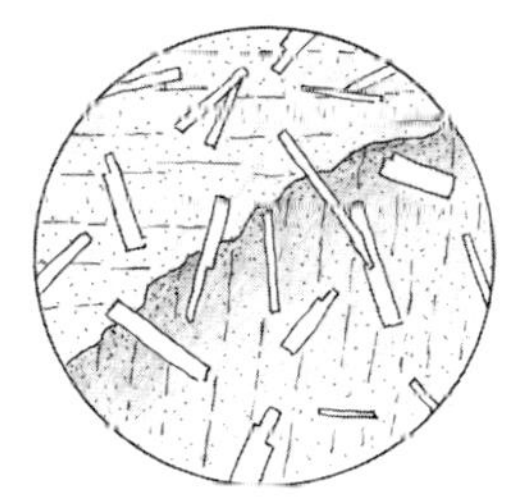

Ophitic texture (*see page 102*) *Plagioclase crystals have become enclosed by the pyroxene augite, as commonly seen in dolerites.*

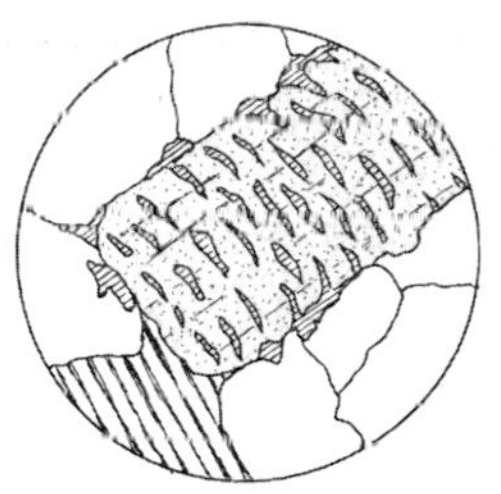

Perthitic texture (*see page 60*) *Crystals of alkali and plagioclase feldspar have intergrown to form perthite.*

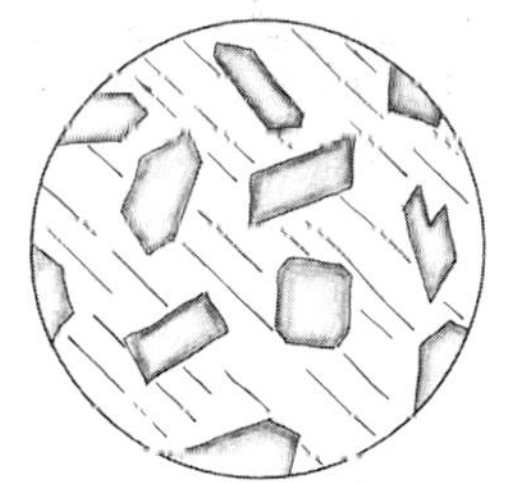

Poikilitic texture (*see page 88*) *One type of crystal partially or wholly encloses several crystals of a different type.*

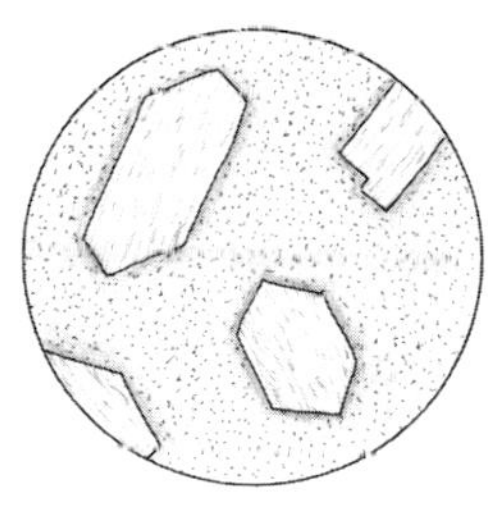

Porphyritic texture (*see page 70*) *Large crystals surrounded by a groundmass of much smaller crystals. A commonly-found texture in many igneous rocks.*

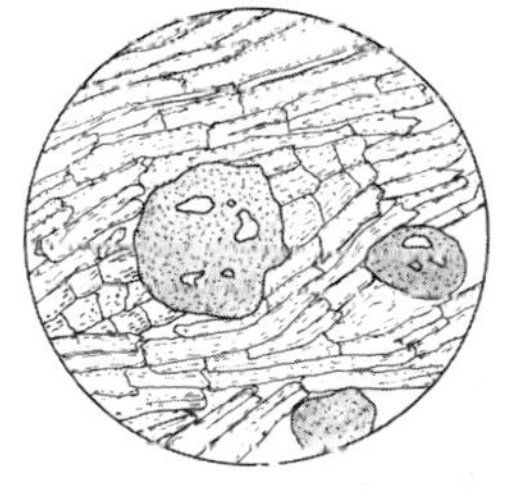

Porphyroblastic texture (*see page 154*) *During the metamorphic process large crystals, or porphyroblasts, grow surrounded by a fine groundmass.*

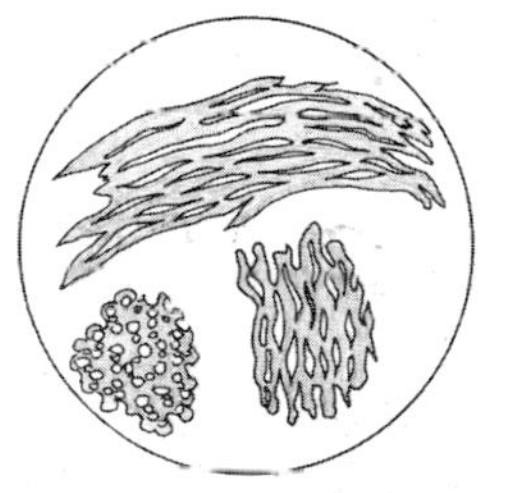

Pumice fragments (*see page 76*) *Sections along (top) and across pumice show its vesicular structure.*

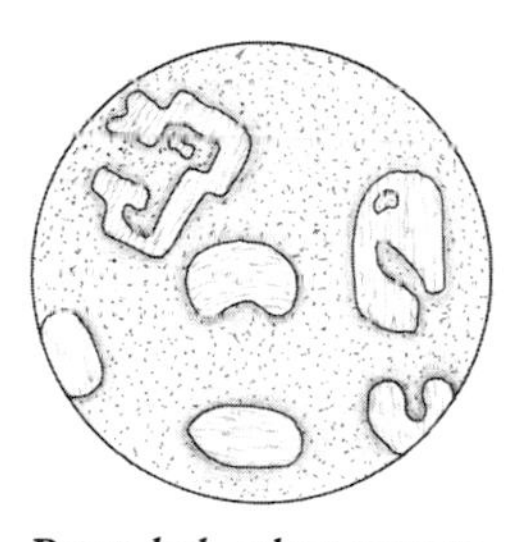

Rounded phenocrysts (*see page 92*) *Phenocrysts become rounded and eroded as they react with molten magma and embayments (indentations) can be formed.*

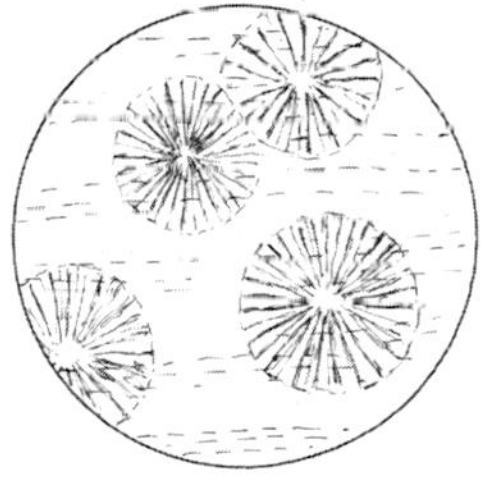

Spherulitic texture (*see page 72*) *Spherical bodies of radially-arranged needle-like crystals often present in rhyolites.*

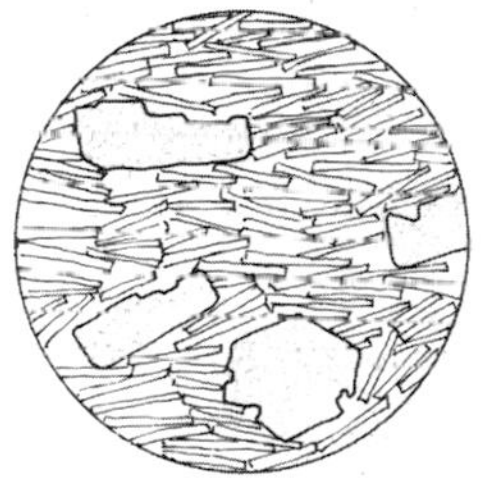

Trachytic texture (*see page 86*) *Feldspar crystals can be seen in roughly parallel alignment. This texture can occur in igneous rocks other than trachyte.*

Index

Page numbers in italics refer to illustrations